Going Digital!

R

Going Digital!

A Guide to Policy in the Digital Age

Robert E. Litan and
William Niskanen

Brookings Institution Press and Cato Institute
Washington, D. C.

Copyright © 1998

Cato Institute

Fax: 202-797-6004

Library of Congress Cataloging in-Publication Data

Litan, Robert E., 1950–
Going digital : a guide to policy in the digital age / Robert E. Litan and
    William Niskanen.
        p. cm.
    Includes bibliographical references and index.
    ISBN 0-8157-5285-7 (pbk. : alk. paper)
    1. Electronic commerce. 2. Digital electronics. 3. Internet
(Computer network) I. Niskanen, William A., 1933– . II. Title.
HF5548.32.L58   1998
651.8—dc21   97-45317
                                                            CIP

                         9 8 7 6 5 4 3 2 1

The paper used in this publication meets the minimum requirements of the American National Standard for Information Sciences—Permanence of Paper for Printed Library Materials: ANSI Z39.48-1984

                        Typeset in Palatino

                    Composition by Cynthia Stock
                    Silver Spring, Maryland

                Printed by Kirby Lithographic Company
                        Arlington, Virginia

Foreword

We may now be at the dawn of the third industrial revolution, in this case one based on the dramatic reduction of the cost of storing, retrieving, processing, and transmitting information made possible by digital technology. The potential effect of this technology is a virtual revolution in the way we work, how we shop, where we live, our health care, our entertainment, our cultural attitudes, and even our politics.

This book addresses the major policy implications of the digital age, with full recognition that it is neither the first nor the last work on this subject. The authors conclude that the government should, for the most part, stay out of the way or, in some cases, actively move out of the way. It is especially important to avoid the pattern of previous policy responses to new technology, where a combination of bureaucratic imperialism and threatened incumbents led to the comprehensive price and entry regulation of several industries. The authors describe a set of conditions for which the government would no longer have a set of policies specific to telecommunications or the Federal Communications Commission (FCC) to enforce such policies. The primary transition responsibilities of the FCC would be to resolve the issues affecting access to the local telephone exchanges and to create property rights in the complete frequency spectrum. The major continuing responsibility of the government is to shape a legal framework that fosters competitive private markets in information technology.

The authors of this book are economists experienced in public policy research and in government service. Robert Litan is director of the Economic Studies program at the Brookings Institution and a former official in the Clinton administration. William Niskanen is the chairman of the Cato Institute and a former official in the Reagan administration. The perspectives reflected in this book were strongly influenced by an April 1997 conference on "Regulation in the Digital Age," organized by the Brookings Institution and the Cato Institute. This stimulating conference involved presentations by and discussions among an eclectic group of economists, lawyers, and information systems specialists, but the views expressed in this book should not be attributed to any member of the diverse group at this conference. (The conference participants and the topics they discussed are shown at the end of the book.) As with all of our publications, the views expressed in this book are those of the authors and should not be attributed to the trustees, officers, or other staff members of the Brookings Institution or the Cato Institute. Readers should be aware that Robert Litan is an officer and shareholder in a company specializing in internet payments.

This book, and the conference that informed it, is the first joint venture by the Bookings Institution and the Cato Institute. Our two institutions have often had somewhat different perspectives on the appropriate role of government but share a commitment to policy debate based on informed analysis and civil discourse. We look forward to addressing selective future issues by such a productive joint venture.

We are especially thankful for the generous support of the April conference and this book by the following companies: Compaq, Data General, Hewlett-Packard, NCR, Silicon Graphics, Sun Microsystems, Tandem, and Unisys. The officers of these companies may not agree with everything in this book, but we hope they share our sense that the book has made an important contribution to the public debate on the policy implications of the digital age.

The authors are especially grateful for the comments and counsel of the following specialists in these issues: Jonathan Band, Solveig Singleton, Robert W. Crandall, Kenneth W. Dam, Austin Fitts, Steve Oksala, Peter K. Pitsch, Jeffrey H. Rohlfs, Alan Schwartz, Osceola F. Thomas, and Joan Winston. Anne Branscomb, who participated in the conference, died in late 1997. The authors are also grateful to Anita G. Whitlock for preparing the manuscript for publication, to Debbie Hardin for editing it, and to Robert Elwood for preparing the index.

MICHAEL H. ARMACOST
President
The Brookings Institution

EDWARD H. CRANE
President
Cato Institute

Contents

1

Introduction

So MUCH HAS BEEN WRITTEN about the information revolution, the information superhighway, the global information infrastructure, and other buzzwords of the current "digital age" that it is easy to become jaded about the technology itself and its implications for how we live. But without doubt, a major technological revolution is taking place, one that may be as important as the agricultural and industrial revolutions before it.

On the surface, digitization seems a rather unimpressive, even retrograde, development. Telephones, radios, and televisions originally all were built using analog technology, by which voices and pictures are transmitted in waves through the electromagnetic spectrum. Waves are continuous—freely flowing. The digital world is the very opposite—a discontinuous world in which any type of information or data (voices, pictures, numbers, and letters) is reduced to bits, or strings of just zeros and ones.

Yet it is the drastic simplification of all forms of data in this manner that has given rise to the enormous advances in technology today. Digitization has made it possible to store, transmit, and receive vast quantities of data, virtually instantaneously, to and from anywhere in the world. Indeed, most computers—including the very first one built with vacuum tubes in 1946, the transistor-based mainframes and minicomputers that dominated computing for much of the post-World War II era, and now the personal computer built with increasingly powerful silicon chips—have long rested on a digital foundation.

The Internet, with its exponential expansion into homes and businesses not only in the United States but throughout the world, is a powerful symbol for the digital age. But in a fundamental sense, the U.S. economy is increasingly going digital. Purchases of computing and telecommunications equipment alone—not counting software—in 1996 accounted for 40 percent of all business equipment investment. By one estimate, the economic activity flowing directly and indirectly from the digital economy eventually will account for half of the output in the industrialized world.[1] Moreover, as firms throughout the economy make increasing use of the new computer and telecommunications technologies and infrastructure to reduce their costs and to develop new products and services, consumers will benefit from lower prices and more choices.

The digital age has broad social and political dimensions as well. In his ominous novel *1984*, George Orwell foresaw some of the technological developments that have since occurred, such as two-way telescreens that we now know as videoconferencing or the speak–write machine that transcribes speech into electronic text, software for which is already on the market (the opposite, the Kurzweil machine that translates text into voice to aid individuals with sight impairment, has already been on the market for some time). But Orwell was wrong, at least so far, in claiming that the new technologies would greatly augment the power of the state and lead to a massive centralization of power. To the contrary, the digital revolution has been a major decentralizing force that has empowered people to change their lives, their institutions, and their governments in previously unimaginable ways. The Berlin Wall came down, after all, not because it was demolished by tanks or armies from the West but because western television and radio coupled with the diffusion of computer technology robbed communist governments of their information monopolies and thus their ability to hold the allegiance of their peoples. It should be only a matter of time when the same thing happens to China (not to mention Cuba and North Korea).

The digital revolution is already stimulating powerful, albeit less dramatic, decentralizing developments in the United States.

---

1. Dertouzos (1996).

As late as 1980, the most advanced U.S. workplaces connected their "knowledge workers" to massive, remote mainframe computers with "dumb" terminals. Today it is difficult to enter any firm and not find workers using personal computers, often networked with each other but capable of operating independently. Decentralized computing and communications, in turn, have empowered citizens in their economic, educational, social, and political activities in ways that would have been hard to imagine twenty years ago.

Access to cheap computing power and to vast databases through local telephone calls to an Internet service provider have reduced the costs of doing business, unleashing a whole new wave of entrepreneurial activity and making it possible for millions of people to telecommute from home. With an ever expanding array of educational software hitting the market, formal classrooms are no longer the only places to learn skills—and this includes both primary education and higher education. Meanwhile, the Internet has erased distance as a barrier to the formation of groups and alliances among people with common interests. Each of these trends is almost sure to continue and most likely gather force in the years to come.

At the same time, as with all technological developments, the digital age has produced its share of discomfort for people and institutions used to living in an analog world. Those of us educated in a world without the personal computer have been forced to learn how to use it or else fall behind our more technocratically facile counterparts and also behind members of the younger generation for whom the PC is taken for granted. Recent economic research has shown that, controlling for all other factors, workers proficient in computer skills outearn other workers by at least 10 percent.[2]

More discomforting to many, the digital age has destroyed not just public information monopolies maintained by governments but also private information monopolies—those monopolies once held by hundreds of thousands, if not millions, of middle managers in their corporations. The brutal result has been corporate downsizing; senior corporate officers find that they can just as

_____

2. Krueger (1993).

easily and more quickly get the information they need directly from their line employees through "intranets" or people using the Internet outside the firm than from now expendable middle managers.

There may be other potential downsides to the continued diffusion of computer and telecommunications technologies that we explore later. But it is vital to keep the negatives in perspective. Technological advances are inherently disruptive. So is competition, which the legendary economist Joseph Schumpeter called the process of "creative destruction." But only a recluse would reverse the clock and do without electricity, the telephone, the automobile, and the transistor—all products of a competitive, market economy.

Moreover, the same technologies that are producing dislocation for some also are providing for many the keys to new opportunities. By dramatically reducing the cost of acquiring, storing, and transmitting information, the digital revolution has made it much easier for millions of Americans, many of them parents seeking to juggle interesting careers with raising their children, to start new businesses. Indeed, many readers of this book are likely to have teenagers who have found ways to make money using their computer skills.

A central challenge for our entire society, therefore, is to find ways of facilitating, if not speeding up, the development and diffusion of digital technologies while avoiding the dangers some believe they pose. That is also our purpose. We were stimulated to write this book by a conference our two institutions convened in April 1997 on Regulation in the Digital Age, at which members of an eclectic group of experts from the fields of economics, law, and business (listed in the appendix) provided their views. We drew much wisdom from their presentations and from the subsequent discussions, but the views we express here are purely our own.

We believe a useful way to illustrate the nature of the challenge just posed is to begin in the next two sections by outlining two alternative scenarios: digital optimism and digital pessimism. The optimistic scenario sketches some of the promise we see from continuing advances in computer and telecommunications technologies, as well as from further diffusion of existing technologies throughout the population. A key theme of the optimistic scenario is that speeding up the diffusion of digital technologies—notably the use of computers and the Internet—is to be valued not for its own sake but for the benefits it can bring to all users, including nondigital businesses and consumers. The scenario is optimistic because it presumes that many of the potential road-blocks to electronic commerce and other digital developments will be overcome, but at the same time it is not euphoric because we do not presume that the projected changes will occur overnight.

We admit that it may be somewhat presumptuous for two nontechies to forecast some of the ways in which our society and economy may benefit from continuing technological advances, just as those who lived in earlier times could not foresee all of the economic and social ramifications of the steam engine or auto-mobile. As one small example, consider that even one of the most important forces behind the digital age, the Microsoft Corpora-tion, was almost eclipsed by the rapid growth of the Internet that the Microsoft leaders did not foresee or take seriously until the company's founder and chairman Bill Gates turned the firm up-side down to embrace the Internet and design products to facili-tate its use. In turn, Microsoft owes its early success to IBM's failure to foresee that PCs were the wave of the future, not the main-frames Big Blue was accustomed to designing and selling. More recently, the buzz in telecommunications circles was about how cable and telephone providers each would provide 500-channel televisions and interactive programming (including movies) to millions of customers across America—none of which has yet hap-pened (although Microsoft's investment in Comcast, one of the nation's leading cable television companies, and its purchase of WebTV may help bring this about).

Notwithstanding these colossal errors in prediction, there has been a sufficient degree of informed speculation about what the future might hold from a broad spectrum of experts in the field that we believe it is possible, by drawing on their writings and using good, common sense, to outline a plausible picture of what a digital future may hold—if not immediately, then eventually.

The pessimistic scenario, in contrast, is easier to illustrate, if only because many of the worries about the digital age have al-ready been highly publicized. Actually, there are two strands of pessimism that we describe. One version is skeptical about the diffusion of digital technology itself, especially of electronic commerce. For example, consumers, fearing a loss of privacy or security, may not use the Internet to make purchases. Or firms, fearing an inability to protect valuable intellectual property, may not make full use of it either.

The other, truly pessimistic version assumes that these problems will be overcome but then fears the consequences of an increasingly digital society for a wide variety of reasons—that it will lead to, among other things, a proliferation of controversial content (pornography, information facilitating criminal or terrorist activities, and so forth), so-called information overload, additional suburbanization, a breakdown of tax collection, and even widespread unemployment. Certain of these fears are exaggerated; others will be the reality to which society will have to adjust, just as citizens have had to adjust to many earlier innovations.

In the final sections of the book we outline the policy principles that, in our view, would best help realize the promise of the digital age without suffering the pitfalls. We are not the first to undertake this task, nor, we suspect, will we be the last. The Clinton administration published a final version of its policy principles on electronic commerce in July 1997, after receiving comments from the public over the preceding six months.[3] The European Union released its version of policy principles in April 1997. The suggested guidelines summarized in this book are broadly similar to those announced by the administration—in particular in strongly urging that the federal government not impose special regulations or taxes on electronic commerce while at the same time encouraging the updating of the legal infrastructure to facilitate on-line commercial activity. Nonetheless, there are important differences in our approach from that outlined by the administration and the EU.

For one thing, our focus is much broader—not just on electronic commerce but on some of the larger social and economic implications of the digital revolution. Moreover, with respect to electronic commerce itself, government must not simply announce a hands-off approach, as the administration has done, but in certain im-

---

3. Clinton and Gore (1997).

portant respects must actively get out of the way by removing legal impediments to interstate and global commerce that were fashioned in a predigital environment. Furthermore, policymakers must remain vigilant against continuing efforts by defenders of old guard technologies and products to stall new advances.

More specifically, we advance in the pages that follow four broad guides to policy in the digital age.

First, policymakers should let markets rather than governments address any problems associated with digital technology. Although the federal government initially financed the development of the Internet, the digital revolution is overwhelmingly a private sector phenomenon and should remain so. Efforts to regulate or tax electronic commerce and other digital technologies may be unenforceable. If they are not enforceable, they run a grave risk of stalling further technological advance, to the detriment of citizens around the world. In fact, the great virtue of a market-based approach is that if digital problems are perceived to be significant by large numbers of users, then firms have strong market incentives to develop technological solutions that premature regulation is likely to preempt.

In fact, the pace of technological change itself favors market-driven over government solutions. Government decisionmaking is slow, whether at the legislative level (where competing interest groups play tugs of war) or at the regulatory level (where rulemaking must be preceded by analysis and accompanied by a notice and comment procedure). By the time government acts, the nature of the problem or the problem itself is likely to change radically. In contrast, corporations increasingly must live or die by Internet time, a reality that tips heavily in favor of market-driven solutions to problems in the digital environment.

Second, governments must recognize and adapt to the shrinkage of their regulatory domains caused by the digital revolution by removing unnecessary and counterproductive barriers to trade and commerce that electronic communications make possible.

To a significant degree this is already occurring. In December 1996, the U.S. government and the governments of many other nations agreed to remove all tariffs on many high-technology products by the year 2000, a step that should unleash even more global trade than now exists in digital products and services. Shortly thereafter, the U.S. government and sixty-eight other member governments of the World Trade Organization agreed to privatize their state-run telephone companies and open up their telecommunications markets to competition. Nonetheless, the governments that were party to the telecommunications agreement must still implement their market-opening measures and, where relevant, privatize their telecom monopolies. Given the history of heavy state involvement in telecommunications, these steps are nothing short of revolutionary. For precisely that reason, the U.S. government will have to monitor closely how these commitments are implemented and, if necessary, apply pressure to ensure that the markets become truly open.

Meanwhile, at home, the regulatory underbrush in a number of areas remains to be cleared. It once was assumed, for example, that doctors, lawyers, and other professionals could deliver their services only by being physically located in the same place as their patients or clients. With videoconferencing and the Internet, that may no longer be the case, provided governments remove outdated licensing rules that prevent out-of-state or foreign providers of these services from entering electronically.

Third, policymakers must resist pleas of incumbent firms and technologies for protection, often under the guise of some motherhood-and-apple-pie justification. Foreign governments that invoke the need to protect their culture against invasions by Americans—and thus to require that all commercial Internet communications be in native language or that some quota of broadcast entertainment be reserved for domestic producers—are doing nothing more than protecting incumbents. The United States is not immune to similar traps. In 1997, for example, the U.S. Supreme Court upheld regulations providing that cable television systems must carry programming by local broadcasters, a ruling that in effect protects only rural broadcasters and prevents producers of newer television channels from reaching the cable audience. More broadly, although federal policymakers are to be commended for abandoning the previous method of assigning specific portions of the electromagnetic spectrum to particular users by auctioning off broader segments of the spectrum instead, the government's continued allocation of spectrum to particular technologies should be abandoned. Although technological limits may dictate that some parts of the spectrum be dedicated to specific uses—such as radio and television—who does what and where on the spectrum should be determined by the market in the form of auctions rather than by bureaucrats.

Fourth, notwithstanding our preference for market solutions, there are certain areas in which affirmative government activity may be useful, if not necessary, to promote digital innovation and diffusion. Furthermore, the benefits of global electronic commerce in particular will not be realized unless some of these required government interventions are carried out on a global scale, not just here in the United States.

One important task for governments is to upgrade the legal and institutional infrastructure to suit the digital age. In a few cases, this may require harmonized rules. For example, as the Clinton administration has pointed out, there is a need to revise the Uniform Commercial Code to permit the use of digital as well as pen-and-paper signatures and to standardize other legal rules relating to commerce in a digital environment. In most other cases, however, harmonization may be either undesirable or impossible to achieve given clear differences in social and political attitudes among nations or states within this country. In those cases, the aim should be to achieve mutual recognition, whereby each jurisdiction agrees to respect the rules and standards of the others. As we will discuss in more depth later, this is the desired outcome in the case of different medical licensing standards within the United States or differences in privacy regimes between this country and the European Union.

At the same time, it is important not to assume that government action may be required to resolve every uncertainty or technological incompatibility associated with electronic commerce. In many if not most cases, markets can do a far better job, more quickly, than can governments. As one example, consider calls for common international rules to determine which jurisdiction's rules—those governing contract interpretation, liability, and consumer protection, to name a few—apply to Internet commerce. Markets may solve such questions all by themselves as sellers make clear in their contracts, as they do now within the United States, which laws apply to their transactions. The history of high-definition television, or HDTV, provides a stark warning to those who would have the government set standards rather than the market. Had the U.S. government heeded the call of those in the late 1980s for the government to set a standard for HDTV, it is almost certain that we would be stuck today with an inferior analog-based technology.

Another way in which government can play a beneficial role in promoting innovation is to ensure that all digital markets remain fully open to competition, because it is competition or the threat of it that has encouraged the development of the digital technologies that we now almost take for granted. Accordingly, there is likely to be a continuing need for some transition period for the Federal Communications Commission (FCC) and state telecommunications regulators to ensure that rivals to existing local telephone monopolies, which carry Internet traffic, have a reasonable chance of competing in the telecommunications business. Regulators and legislators have so far refused to recognize that this objective is inconsistent with efforts to distort the telecommunications rate structure in order to subsidize particular classes of users or to attain so-called universal service. The best way to guarantee universal service is to promote competition, which will lead to lower prices without the implicit subsidies. Eventually, once competition takes hold in local telecommunications markets, the need for continued state and federal regulation, as well as the agencies that now administer it, should disappear.

A similar fate should not be in store for antitrust enforcers, however, because ensuring competition in digital and other markets remains an essential function in a market economy. At the same time, however, the antitrust authorities should take account of the increasing returns that characterize a number of digital technologies and that therefore tend to cause the markets in which those technologies are present to be more concentrated (served by fewer competitors) than other markets. In particular, some digital technologies—notably, the production of microprocessors—require major investments in research and development, which are fixed costs that do not vary with production. Accordingly, as the sales volumes of these technologies expand, the average costs of producing them necessarily falls. Moreover, some high-technology products are characterized by network externalities, which means that they become more valuable as others use them. The telephone is an obvious example, but so is much computer software. Furthermore, digital products can be complicated. Once users invest time and effort to become comfortable using them, they can be very reluctant to switch—familiarity with word processing and spreadsheet programs being obvious examples. For all these reasons, antitrust authorities will have to live with the fact that some digital markets may be characterized by monopoly or near monopoly for some period of time. The main aim of antitrust should not be to punish the victors in such markets but instead to ensure that these markets remain contestable by new entrants offering better technologies (which, in turn, may become new, albeit temporary, monopolies).

It is also critical for antitrust authorities to recognize that the digital age is characterized by many different, related technologies working together—hardware and software, operating systems and applications programs, content producers and broadcasters, and so on. In the language of the field, all of these technologies must be interoperable, which in turn requires the development of standards, or protocols, that allow one technology to work with another. In some cases, standards developed by a single firm can become widely accepted in the marketplace (as in the case of Microsoft's operating systems). In other cases, however, standards can only be developed through collaboration among a number of parties, many of whom may be competitors. The standards for communication over the Internet, for example, have been developed cooperatively. A key lesson for antitrust is to be tolerant of joint ventures that are created to develop and refine standards and certain digital technologies, which have been essential to the digital age so far and almost surely will continue to be so in the future.

2

Digital Optimism

MICROSOFT CHAIR Bill Gates floats the promise of "friction-free capitalism."[1] Technofuturist George Gilder expects that, "The Internet will be the central nervous system of the new global economy."[2] Nicholas Negroponte of the MIT Media Lab projects a billion Internet users worldwide and a trillion dollars of electronic commerce early in the next century.[3] Also at the conference, Louis Rosetto, publisher and editor of *Wired* magazine, described his cybernaut readers as "a new breed: nonpartisan, libertarian, inherently tolerant, politically incorrect, skeptical of the established media, less threatened by government than believing it outdated and irrelevant."[4] These and other cyberoptimists describe a virtual revolution in the way we work, how we shop, where we live, our health care, our entertainment, our cultural attitudes, and even our politics.

Are they smoking pot? Maybe so, but they may still be right. Let us start with what is known. For most Americans, the digital evolution is barely visible. Although about 40 percent of households now have a PC, about 20 percent are connected to the Internet and only 2 percent are regular users. Consumer purchases from the Internet, now less than $1 billion a year, and the use of

---

1. Gates (1995).
2. Brookings-Cato conference, "Regulation in the Digital Age," April 1997.
3. Negroponte (1995).
4. Brookings-Cato conference, "Regulation in the Digital Age," April 1997.

digital cash are still trivial in an $8 trillion economy. Video sex is the largest current retail service on the Internet, gambling is probably a close second, but financial services are growing rapidly—more than $100 billion of assets are now managed on-line.[5]

Some services made possible by digital technology are now more visible. Automatic teller machines (ATMs) now provide routine banking services at more than 120,000 U.S. locations. Some utility companies now use electronic billing, saving two pieces of mail. Most children are enamored of computer games with crude video pictures; they may not be aware that the realistically scary films of dinosaurs and spaceships are also created by computers. Some automobiles now include an electronic navigator. The cellular telephone is rapidly becoming a pager, answering machine, information terminal, and electronic key ring. So far, however, these services are peripheral to the major concerns of most households.

The digital revolution is more apparent in the workplace and in commercial markets. Purchases of information technology are now about one-third of total private investment in producer-durable equipment. PCs are now standard office equipment, and millions of workers now telecommute. More than 80 percent of Fortune 500 firms now have a website. Business-to-business sales overwhelm the on-line sales to retail customers, but the huge business investment in computers, communications, software, and training has not yet increased the growth of (measured) productivity.

So what is the basis for optimism about a digital revolution? Momentum! For the past several decades, the computing power of microprocessors and the number of Internet users have each doubled every year or so. Such growth cannot continue indefinitely, but there is little evidence to date that this process has slowed. And now the absolute increase in computing power and the number of Internet users from each doubling is very large, opening new technological and economic opportunities. The other basis for optimism has been the anticipatory creativity of the cybernauts in their effort to "imagineer" the potential applications of a continued rapid growth of computing power.

---

5. "A Survey of On-line Commerce," The *Economist, May 10–16, 1997, p.11.*

In short, the direction of the effects of digital technology is clear, even if the magnitude of its potential effects is still uncertain. Digital technology dramatically reduces the cost of storing, retrieving, processing, and communicating information. The major specific effect of the Internet is to nearly eliminate the effect of distance on the cost of communication. The potential consequences are enormous. Michael Dertouzos, director of the MIT Laboratory for Computer Science, may be the most realistic of the cyberoptimists. In his new book *What Will Be*, he predicts that the following types of services will be available in the next decade or two:

—A virtual job market will make it possible to find employment opportunities through a specialized on-line broker, conduct on-line job interviews, and telecommute from one's house to a distant firm with only occasional direct contact.

—Virtual shopping arcades will display the available products, provide for a virtual experience of trying on new clothes or driving a new car, and [allow customers to] order a product—all without ever visiting a conventional retail store.

—A virtual health care system will permit remote testing and diagnoses, with direct care provided, possibly, only by a medical technician.[6]

Dertouzos is careful to note that the infrastructure of the information marketplace is not yet sufficient to provide these services and that many future uses of this marketplace are now beyond imagination. (He should also have mentioned that the potential for telemedicine is seriously limited by state licensing laws that restrict the use of out-of-state professionals and the division of labor in the health care system, a subject we discuss later.)

Many of the specific applications of digital technology will continue to be a surprise, but the general direction of the effects on our major institutions is increasingly clear.

---

6. Dertouzos (1996).

The Workplace

Organizations will become flatter. The Internet and organization-specific intranets will reduce the relative number of middle managers and middle staff positions, although the Internet is already creating new jobs for those who are proficient in navigating through it.

For a while, maybe for a generation, the relative earnings and influence of computer-literate employees, often younger employees, will increase.

More employees will telecommute more of the time. Organizations will also subcontract tasks, often to other countries, for which the output can be transmitted on-line.

Teleconferencing will probably reduce business and professional travel.

It is less clear whether digital technology will increase or reduce the economies of firm size. Our judgment is that firms will become more specialized, reducing both horizontal and vertical integration, but that concentration within some specializations may increase.

The old cliché stands: Information is power, for individuals acting alone and for those within an organization. As more information is generated and made available to virtually anyone, inefficiencies and failures are more rapidly exposed, enhancing the role and influence of performance-based management systems.

And, as a rule, governments will be the last organizations to adapt to these effects.

The Marketplace

The primary new effect on the marketplace will be to reduce the relative number of agents: travel agents, real estate agents, securities brokers, auto dealers, other retailers, and so forth. Sellers will find new ways of communicating to potential buyers without someone acting as liaison. Consumers will find new ways of making a market, such as by scanning the options on-line and then asking for bids to sell a specific new car model. This will lead to lower and more uniform prices, by reducing both selling costs and inventories. The elimination of any significant distance effect on the costs of communications, of course, will accelerate the globalization of commerce.

The Home

The home, in contrast to the firm, will be less specialized, serving as the locus of many different activities formerly conducted in different locations. Thus more adults will work at home; more students will study at home, with access to superior on-line instruction and libraries; and a wider variety of entertainment will be available at home. People will have more choice of when they want to participate in each of these activities. And such computerized devices and services as smart stoves and cleaning robots will serve the household.

Culture

The digital revolution, like the recent Pathfinder mission to Mars, is a "revenge of the nerds." The nerds or cybernauts have a clearly distinctive culture, many attributes of which are quite admirable, especially compared with that of some other contemporary cultures. As described by Jon Katz,

> This nascent ideology, fuzzy and difficult to define, suggests a blend of some of the best values rescued from the tired old dogmas—the humanism of liberalism, the economic opportunity of conservatism, plus a strong sense of personal responsibility and a passion for freedom.
>
> Some of their common values are clear: they tend to be libertarian, materialistic, tolerant, rational, technologically adept, disconnected from conventional political organizations . . . and from narrow labels like liberal or conservative. They are not politically correct, rejecting dogma in favor of sorting through issues individually, preferring discussions to platforms.
>
> They share a passion for popular culture—perhaps their most common-shared value, and the one most misperceived and mishandled by politicians and journalists.
>
> Unlike the clucking boomers, they are not talking revolution;

they're making one. This is a culture best judged by what it does, not what it says. [7]

This culture is attractive to many individuals, is threatening to many dogmas, and will be a force for change, at least for a while, but the ultimate effect of this distinctive culture is still quite uncertain. How will it change as the mostly young cybernauts mature? How will those groups threatened by this culture respond? Will the continued digital revolution broaden or dilute this culture? We do not yet know.

But one thing we do know is that the Internet has already given rise to a new form of communication: e-mail, which to some is a curse, but to most others, probably, it is an incredibly welcome development, spurring more frequent and spontaneous communication between families, professional colleagues, friends, and new Internet acquaintances.

Politics

Orwell's ominous vision of the political effects of electronic technology, of course, was very different from that of Katz. Orwell feared that the ubiquitous telescreen would greatly increase the power of the state to monitor and control individual behavior. Katz observed that cybernauts are strongly committed to privacy and individual freedom. Others have speculated that the ready availability of information about our political system, whether on-line or through television, will enable instant, mass referendums on political issues. As in the private sector, the role of intermediaries—in politics, elected representatives—will become less relevant. In short, the same technology has led to drastically different, and to some people more or less equally ominous, visions of our political future.

Is there any sensible way to sort out the effect of new technology on politics, the relation between people and the state? Again, the answer cannot be known for some time, but several observations are worth making now.

---

7. Katz (1997).

First, it is important to avoid embracing too quickly any form of technological determinism. The three technologies associated with the rise of the West in the sixteenth century, for example, were gunpowder, the magnetic compass, and the printing press. But the Chinese had developed each of these technologies a century or more earlier before withdrawing from most foreign contacts.

Second, the differential availability of technology has had a powerful effect on who governs but a much smaller effect on the nature of government. Inventions such as the stirrup, which in retrospect seem so simple, led to the rise and fall of major civilizations without much changing the relations between the average person and the state.

Third, some technologies nonetheless seem to have had general effects on the form of government. Gunpowder contributed to the rise of the nation-state because it reduced the effectiveness of point defense relative to area defense. The nuclear weapon, in contrast, has had a decentralizing effect, because it gives small nations the potential for a devastating attack on much larger nations. The printing press, radio, and television have clearly reduced the power of all types of imperial ideology.

So what can be expected to be the major political effects of digital technology? The essential feature of government is a monopoly of the legal right of coercion in some defined space. Yet the obvious and dominant political effect of digital technology is to reduce the monopoly power of governments, a subject we explore in greater depth later when we take up the issue of taxation in the digital age. The impact on governments throughout the world reaches far beyond their power to tax, however. A well-informed population is not one that can be easily subjugated or fooled. The diffusion of information already has contributed to the downfall of communist regimes in the former Soviet Union and Eastern Europe. Democratic capitalist societies are no less immune to the empowering effects of the information revolution brought about by advances in digital technology.

One other potential effect of the new technology is especially intriguing. Powerful computers and creative software have facilitated a new form of social science: testing the effects of alternative social, economic, and political rules. The new Brookings

Institution book by Joshua Epstein and Robert Axtell, *Growing Artificial Societies: Social Science from the Bottom Up*, for example, uses the computer to simulate social systems.[8] This agent-based modeling technique could well become the dominant form of carrying out social science in the next century, and it may create a common scientific foundation for ethics, economics, and politics. This is an exciting prospect, even for those of us who are low-tech, older social scientists.

A parallel effect of the digital revolution on politics, of course, is the culture of a digital nation. As already described, however, it is not yet clear how much the shared values of this culture will be accepted by the larger community as more extensive use is made of digital technology.

There are some disturbing signs that point in more pessimistic directions, however—which we explore in depth shortly. The digital infrastructure of the Western world promotes freedom because it is decentralized and privately owned. But for most of the developing world, digital networks have yet to be put in place. They could end up with a much different result, owned by the state and used by governments as instruments to restrict rather than expand political freedom.

How will all this turn out? Again, we do not yet know. The issue of whether the digital revolution empowers individuals relative to the state is not written in the stars; instead, this issue will be resolved by the choices that all of us have yet to make.

---

8. Epstein and Axtell (1996).

3

Digital Skeptics and Pessimists

NOT EVERYONE WILL BELIEVE or be enthralled by the scenario just outlined. Those whom we call digital skeptics believe key parts of this scenario will never happen—or at the very least, may not develop any time soon—because legal, institutional, and technical roadblocks are standing in the way. Those whom we label digital pessimists do not doubt that these barriers will be removed but fear what will happen if they are.

Digital Skeptics

Digital skeptics focus their attention primarily on electronic commerce and question whether growing numbers of people will trust doing business anonymously over the Internet with counterparts they cannot see or whose identities they cannot verify. As one famous cartoon in the *New Yorker* has illustrated, you can be a dog on the Internet and no one will know the difference. The skeptics will observe that, despite much media hype, it is generally believed that consumers are purchasing no more than $1 billion per year of goods and services on the Internet—a tiny fraction of the U.S. $7 trillion-plus in annual output—and they will point to a number of barriers, any one of which can frustrate further development of electronic commerce, even if all of the others are removed. Moreover, digital skeptics might properly note, perceptions about the importance of each of these impediments can be, and probably are, more important than the reality: as long as sizable numbers of people believe any one or more of the problems are significant, they will be deterred from using the Internet to conduct business regardless of the objective circumstances.

Security

Perhaps the most significant reason why there is not more business conducted electronically is that potential purchasers fear that their communications over the Internet—in particular, their credit card or bank account numbers, medical information, personnel records, and trails of which websites they may have visited—can be easily intercepted by those to whom the messages are not addressed. To be sure, some danger now exists when people order goods over the telephone and provide their credit card numbers to clerks they have never met, or even when people use their credit cards to pay for goods and services in person (because the numbers on the cards can easily be copied by one or more individuals who handle the processing of credit card slips as they proceed from the retailer through the banking system).

The skeptics argue, however, that the security problem on the Internet potentially is of a different order of magnitude, for at least two reasons: the range of information that might be available is much broader and so is the universe of people that can potentially access it (theoretically, the tens if not hundreds of millions of users of the Internet). Moreover, there is the problem of trust: even if consumers have confidence in the technical capabilities of credit card companies to safeguard account numbers, they have no way of knowing for sure (other than by recognizing brand names) whether the companies from which they may be ordering merchandise over the Internet are legitimate enterprises or fraudulent operators.

In fact, concerns about computer-related security extend far beyond the realm of electronic commerce, to the real world of today, where vast amounts of critical confidential information are stored in computer databases and often transmitted through the nation's telecommunications system. Consider, for example, the potentially devastating consequences if criminals or terrorists cracked one of the existing national information systems, such as the air traffic control system or the Fedwire system, which the Federal Reserve System uses to transfer more than $1 trillion every day between banks. As it is, hackers already have penetrated the computer systems of many of the nation's largest corporations and even of the Defense Department, apparently on a regular basis.[1] In October 1997, members of the President's Commission on Critical Infrastructure delivered a classified report to the president on the susceptibility of the federal government's computers in particular to sabotage and reportedly offered a series of recommendations to counter the threat.

The generally recognized solution to computer security problems is to encrypt messages or data containing confidential information into a lot of numbers and letters, which can only be decoded by those who have the keys or the algorithms that created the ciphers. The strength of the encryption, or how easy it is to crack, depends on the robustness of the algorithm used and on the length of the string of characters used in the key, measured in bits. Currently, weak encryption uses forty-bit keys; stronger encryption uses fifty-six bit keys that are more than 65,000 times harder to attack. Most banks now use 128-bit or even stronger security, and a U.S. firm has produced (in Japan) a chip that provides 1,024-bit security. As it is, encryption techniques are widely used by governments around the world to protect military secrets, by private companies to protect confidential information, and by financial institutions as they exchange payment and other sensitive data.

Encryption has its limits, however. In the right hands, it is a powerful tool for protecting the confidentiality of information and preventing it from being misused or stolen. But in the wrong hands, encryption can make it easier for criminals to conduct their activities without fear that law enforcement authorities will be able to successfully prosecute them; indeed the commercial availability of such encryption algorithms as PGP (an acronym for pretty good privacy) has already made it easier for criminals to do business over the Internet. The tensions over encryption policy have led to a major policy dispute between the Clinton adminis-

---

1. Dam (1996, p. 9).

tration and the high-technology community, especially as it re-lates to the exportation of encryption software, which we discuss later.

Encryption is limited in other respects as well. As Professor Dorothy Denning pointed out at the Brookings-Cato conference, cryptography does not protect systems from viruses or electronic time bombs that can be innocently downloaded from the Internet with instructions to destroy files or transfer funds from one account to another. In addition, even the strongest encryption in the world is of little value if hackers can find ways to subvert a user's operating system before he or she even sends an encrypted message.

In short, digital skeptics can argue that computer-based systems for storing and transmitting information will remain susceptible to misuse. At the very least, the skeptics might suggest that continued suspicions among the public about security will deter use of the Internet for business purposes. At the worst, some national information system will suffer a serious electronic disaster, which not only could have immediate devastating consequences for the users of the particular network that may be affected but also could cast a long shadow over the rate at which digital technologies are used across the board.[2] In fact, in June 1997, a computer technician successfully used the Internet to organize 14,000 computers in a brute attack to "guess" a 56-bit Data Encryption Standard (DES) key. DES is widely used by financial institutions and other companies to protect their sensitive information. This event, skeptics argue, demonstrates the fragility of security not only over the Internet but in electronic media generally.[3]

---

2. Deutch (1996).

3. Optimists would counter that despite the successful attack on a single DES key, U.S. banks are still quite secure because they use additional means (such as changing encryption keys regularly) to strengthen their systems. The Federal Reserve, in particular, is moving quickly to implement a more effective mode of DES with a longer encryption key, making it far more difficult to attack than the DES-encrypted message that was broken in June 1997.

Privacy

Concerns about privacy on the Internet clearly are related to security: if users' communications are not secure, then they are exposed to the risk that information they believe to be confidential could be obtained by others without their consent. The Internet also has aroused concerns that even the parties with whom users intend to communicate can invade their privacy either by using the data they may supply over the Internet for other purposes or by selling the data to other parties without their knowledge (let alone consent). To make matters worse, personal data that may be circulating on the Internet could be wrong, damaging reputations in the process.

Security breaches contribute to privacy concerns on the Internet. When the Social Security Administration announced during 1997 that it would make available individual-specific benefit information on the Internet, it was quickly forced to back down over concerns that the service would not be secure.[4] Similarly, in August 1997, when Experian—one of the nation's largest credit bureaus—made available to individuals on the Internet their credit information so they could promptly correct it if it was in error, it was forced to withdraw the service the next day when certain individuals got access to the wrong credit histories and that fact was publicized by the media.

In fact, concerns about privacy extend well beyond the Internet. Americans have been supplying information about themselves to a wide variety of recipients for years, well before the Internet was born: when we apply for credit cards or frequent flyer programs or when we file for bankruptcy or divorce (where records are open to the public), to name but a few examples. Many firms in the nondigital economy also have been selling subscription or mailing lists that contain the names and addresses of countless numbers of individuals. And there have been celebrated instances of abuse. In June 1997 a *New York Times* article elaborately de-

---

4. The agency has since offered a more limited service whereby individuals with registered e-mail addresses may be able, after clearing several security hurdles, to obtain limited benefit information. Those who want full information can use e-mail to request that it to be mailed to them.

tailed the story of a woman who was stalked from prison by an inmate who, while working for a contractor to a large retail company, gathered personal information from responses to the company's product questionnaires. The Internal Revenue Service, meanwhile, has been embarrassed by disclosure that some of its employees have been wrongfully examining tax returns of individual taxpayers.

Digital skeptics argue that because it so dramatically cuts the cost and time of accessing information, the Internet poses privacy problems to a much greater degree than was previously thought possible, and for this reason will stymie further growth of electronic commerce. Many in the public seem to agree. National surveys conducted by Louis Harris and Associates report that 85 percent of Americans say they are concerned about threats to their personal privacy.[5] Harris polls also indicate that among those who have not yet used the Internet, fear about privacy is the single most important reason why they have not done so.[6] Even among users of the Internet, more people (more than half) are more concerned about the confidentiality of their communications over e-mail than over any other form of communication. The subject of privacy in the digital age—and fears that citizens are losing it—has become so topical that it was featured as a cover story in a recent edition of *Time* magazine.[7]

Skeptics can also point to another potential impediment not only to the use of electronic commerce but to data transmission more broadly: the Privacy Directive that will become effective in the European Union in 1998. Under this directive, the EU will decide whether the privacy protections offered by other nations are adequate; if they are found to be inadequate, then the EU will prohibit all transfers of all personal information about their citizens to countries that do not meet this test unless those who propose to send data qualify under certain exceptions. A looming question is what the EU will decide about the privacy regime in the United States, which consists of a blend of federal and state statutes and case law but not the kind of comprehensive system

---

5. Reported in Westin (1997).
6. *Wall Street Journal*, June 19, 1997, p. B6.
7. Quittner (1997).

that exists in some European countries that require gatherers of data to register with government privacy offices. In addition, all European countries prohibit or impose limits on various data uses (such as those for direct marketing) that are routine here.

Continued insistence by the EU that the United States adopt a very similar system, which has virtually no support in this country, could lead to major trade frictions. If the EU carries through with its threat to embargo personal data going out of Europe, it will not only harm U.S. companies now doing business in Europe as well as European consumers themselves, but it could significantly disrupt electronic commerce.[8]

Taxation

In principle, electronic commerce could be nipped in the bud if jurisdictional entities within countries (state, county, or city governments) or national governments see it as a potential source of additional revenue and begin to place special taxes on transactions completed over the Internet. In the strongest possible terms, the Clinton administration has urged that no new taxes or cross-border tariffs be placed on electronic commerce. Legislation sponsored by Senator Ron Wyden and Representative Chris Cox would ensure that this policy is carried out at least within the United States by imposing a moratorium on new state and local Internet-based taxes on sales and services.

Digital skeptics fear that these well-intentioned efforts could easily be derailed by jurisdictions both within and outside this country intent on gathering as many golden eggs as they can from the goose of electronic commerce. For example, they might note that the Wyden-Cox proposal is being opposed by many state and local governments, some of which have already imposed sales and use taxes on Internet access charges and on sales of goods downloaded from the Internet.[9] Some countries may be tempted

8. For a more extensive treatment of the issues raised by the European Privacy Directive, see Swire and Litan (forthcoming).

9. Some states, however, have endorsed Wyden–Cox or similar proposals, including the California State Board of Equalization and the states of Massachusetts and New York. In addition, some states have already adopted

to do the same or recover even greater revenue, notwithstanding the U.S. government's urgings.

Whether special Internet-based taxes are assessed on sellers or buyers or not, we believe over the long run they are not likely to raise much revenue, and for that reason many jurisdictions are unlikely either to implement them or, if they do, maintain them for lengthy periods or set them at onerous levels. Sellers of goods and services over the Internet are highly mobile. If subject to special taxes, they are likely simply to move the location from which they offer their wares. Although buyers are not so mobile, putting taxes at their end of the electronic pipeline would just encourage them to continue using more conventional means (the telephone, the mail, or in-person visits) to complete their transactions.

Although the skeptics almost surely will be wrong about Internet taxes, the growth of electronic commerce has significant implications for the assessment and collection of current conventional taxes, such as those on income or sales, regardless of how they are generated. We take up this issue shortly.

Intellectual Property Protection

Another commonly cited barrier to the growth of electronic commerce is the absence of clear intellectual property—principally copyright—protection for innovative content displayed on the Internet. This issue was not addressed at the Brookings-Cato conference and thus our views here are more tentative than those we express on other subjects. Nonetheless, certain propositions in this area seem defensible.

Through decades of court decisions, the application of copyright law to paper-based content—books, newspapers, and magazines—has become relatively clear. The law broadly protects creators and publishers of original content from unlicensed uses of their works so that they have strong economic incentives to produce them. At the same time, copyright gives users rights of "fair use" to make copies of copyrighted materials for private,

---

or, at this writing, are considering, moratoriums on Internet taxes. For a current listing of Internet-related taxes, see Vertex at http://www.vertexinc.com.

noncommercial purposes. The copier machine, for example, could not exist without the fair use doctrine.

The arrival of the digital age has triggered a debate about how, if at all, to update copyright law to suit the unique characteristics of cyberspace. One set of digital skeptics (and advocates for stronger copyright protection) argues that because electronic images can be easily copied—without charge—and retransmitted to millions of users around the world, paper-based copyright law must be updated by the Congress and other governments (through the World Intellectual Property Organization, WIPO). One key suggestion is that content originators be given rights in the temporary reproduction in computer memories of original material transmitted in cyberspace, which would mean that every time users browse on the Internet, they must obtain a license even to download material temporarily (under current law, making permanent copies clearly requires a license). Without these modifications of copyright, some skeptics argue, the Internet will never realize its full promise because the best content originators will avoid it.

To a significant degree, such warnings are unduly alarmist. The absence of stronger copyright protection has not prevented a growing number of web-based publications (such as *Slate* and *Hot Wired*) from offering their content on the Internet, nor has it inhibited an even more rapidly growing number of firms from offering their products and services the same way. Indeed, adopting the measures just outlined could severely damage the growth in electronic commerce and the use of the Internet that has already occurred. Assuming it could be enforced, a requirement that users obtain licenses simply to download information from the myriad web pages now available would be inconsistent with the free-flowing ethic of web-surfing that has made the Internet so popular. In combination, all of the foregoing measures would severely curtail use of the Internet, thus killing the very technology of which content originators presumably are seeking to take advantage.

This is not to say that the existing paper-based copyright system is adequate to protect many forms of entertainment, such as movies or sound recordings, that may be transmitted over the Internet. It is not. But then again most users currently do not yet have sufficient bandwidth in their Internet connections to download such data-intensive transmissions, which affords some time for policymakers and industry to address the problem. The solution may not lie in the realm of the law but in technology. Just as movies on rental videocassettes cannot be copied legally, we suspect ways will be found of encrypting copyrighted forms of entertainment to prevent users who lease them for limited viewing or listening over the Internet from retransmitting the works to other parties. In the meantime, "watermarking software" already has been developed to allow content originators to track where the images they have created appear elsewhere on-line, a technology that makes it possible to track down Internet pirates.[10]

At the same time, certain features of the existing paper-based copyright regime seem ill-suited to the digital age. The period of a copyright, now—for the most part—the life of the author plus fifty years, makes little sense in a world where the economic half-life of software or popular music may be around eighteen months. Some shorter copyright period in the digital age may be appropriate.[11]

Treaties on copyright reform adopted by WIPO, which at this writing are pending before Congress, attempt to grapple with some of the issues just discussed. However, the proposed legislation (S. 1121 and H.R. 2281) that would implement the treaties, unlike the treaties themselves, would target the use of specific technologies rather than specific illegal behavior. We believe this is a mistake. As we discuss in connection with content control, it is important to apply the same rules to enforce copyrights, regardless of the means by which a person violates the rules. The domestic information technology industry has suggested alternative technologically neutral language that, among other things, links copyright liability to an intent to infringe. Congress should give serious attention to this proposal.

---

10. Similar watermarking technologies are being developed for music distributed online.

11. To those who might claim it unnecessary to shorten the copyright life of digital works because the market seems to be reducing the economic value of such material, we would point out that the lengthy copyright period nonetheless casts a long legal shadow that can give rise to unnecessary litigation. Better simply to align the law with economic reality.

Furthermore, although the proposed copyright legislation for the most part makes only modest changes in copyright law, its general prohibition of technologies aimed at circumventing technologies that control access to a copyrighted work is problematic. In particular, the proposed language threatens efforts to enhance security, privacy, and parents' efforts to limit access by their children to unwelcome content on the Internet:

—The bill could chill encryption research because it appears to allow researchers to crack encrypted security systems only if content owners give their permission to do so. Cryptography progresses, however, through the freedom of researchers to crack encrypted works. Requiring advance permission for this activity to proceed will slow and probably dampen research in this important area.

—The anticircumvention language in the bill would restrict an individual's ability to protect his or her privacy by disabling programs that (often unknown to computer users) transmit information to other parties that reside on a user's hard disk. Such "cookie" programs (which we discuss later) are copyrighted works.

—The same anticircumvention language would make it impossible for parents to use programs to review what pages their children viewed on the Net if the sites used anticopying technology to prevent that kind of monitoring.

Totally apart from the debate over the WIPO treaty and its implementing legislation, on-line service providers argue, with justification, that the law currently exposes them to an uncertain degree of liability for copyright infringement by users of their services. These service providers are akin to the postal service or private mail carriers and should not be chilled from sending content over the Internet. A clarification of copyright law to ensure that on-line providers are not made liable for subsequent copying (unless they have control over the content of the communication or message) would be in the interest of content originators and the public alike.

Customer Acceptance

Finally, even if all of the foregoing technical and legal problems are resolved, some skeptics nevertheless might question the rate at which customers actually will use computer-based technologies to make purchases, pointing to the following facts: that despite all the hype about the computer revolution, as we have already noted, roughly 40 percent of American families have a PC and only about half of them are hooked up to the Internet. Perhaps most significant, as of early 1997, only about 2 percent of American households reported that they made "heavy use" of the Internet, compared with a 98-percent penetration rate for televisions and radio, 89 percent for VCRs, 34 percent for cellular phones, and 19 percent for modems or fax machines.[12]

Skeptics can cite other evidence as well to question the pace at which electronic commerce will grow. As long as thirty years ago, optimists were forecasting the end of the paper check, to be replaced by electronic communications. Today, even with credit cards, ATMs, telephone and computer banking, and all the talk about electronic bill payment, checks remain the workhorse of the payments system. This experience, skeptics will claim, should humble the digital optimists who only see rapid growth in electronic commerce ahead. And as for consumers actually buying things over the Internet, skeptics will argue that all but hard-core technophiliacs will want to see and touch their merchandise before buying it—not to mention those who want to enjoy for its own sake the experience of getting out of the house and shopping in person.

Digital Pessimism

As we argue in the following section, we believe that each of the hurdles to the diffusion of electronic commerce can be resolved or overcome, if not primarily by market forces, then by affirmative government action. Moreover, we are not convinced that if the barriers to electronic commerce truly come down, consumers will be as resistant to using the vast capabilities of the Internet to conduct business as the skeptics believe. Notwithstanding all of the statistics just cited, the digital revolution is in its incipiency, and its hardware is penetrating the population more rapidly than earlier innovations.

---

12. *Wall Street Journal*, June 16, 1997, p. R14.

For example, one useful measure of diffusion is how many years it has taken for a technology to have penetrated 25 percent of households or other relevant units. As benchmarks, electricity, the automobile, and airplane travel each took about half a century to reach the 25-percent penetration mark, whereas radio and television took about a quarter century. In contrast, it has taken fifteen years for the personal computer to reach that penetration rate, and cellular phones just thirteen years.[13] As microprocessors become ever more powerful, computer-related costs will continue to fall, bringing the technology within reach of an expanding fraction of American households, and even more attractive and useful applications and content will continue to be developed. Internet use will surely increase in their wake.

Ironically, none of this is reassuring to those whom we call true digital pessimists, those who see a variety of dangers rather than promise not only from wider use of the Internet but from the use of digital technologies more broadly. Before we examine some of the claims in more detail, it is important to retain some perspective. All new technologies, by definition, threaten the existing order and in the process unsettle those accustomed to living with the status quo. Without exception, technological advances also lead to unanticipated outcomes, some unexpectedly beneficial, others not so desirable. The automobile, for example, has given unprecedented mobility to billions of people around the globe; but it also has created urban sprawl, air pollution, and traffic fatalities. The invention of nuclear power helped lead to cheaper electricity, but also has left the nation grappling with how to dispose of its highly radioactive nuclear waste. The digital age is not likely to be an exception.

A critical question for policymakers is whether any of the purported dangers are sufficiently likely and serious that some sort of preemptive, or even after-the-fact, intervention by government is appropriate. We generally believe not. Some of the fears we outline next seem to lack any reasonable basis. As for some of the others, our counsel is that governments should avoid overreacting to alarmists' claims, because the march of the technology itself should re-

---

13. *Wall Street Journal*, June 16, 1997, p. R4.

solve many of them. And for the remainder, our broad advice is for society simply to get used to living in a digital environment, just as it has adapted to so many previous innovations.

Controversial Content

Perhaps the best-known complaint about the Internet is that it affords much too easy access, especially to children, to pornographic and other controversial material (such as instructions on bomb-making and other information of use to criminals and terrorists). Governments have not hesitated to act to meet these concerns, which seem to be shared by large numbers, if not ample majorities, of the citizens in most countries.

For example, Congress included in its sweeping overhaul of telecommunications regulation in 1996 a provision known as the Communications Decency Act (CDA) that subjected to criminal prosecution anyone who displays "indecent" or "patently offensive" material on the Internet in a manner that would be available to those under eighteen years old. Content providers could avoid prosecution if they required the use of a credit card or other "reasonable, effective, and appropriate actions" to prevent access by minors. Shortly thereafter, the CDA was struck down as an unconstitutionally over-broad infringement on free speech by a federal district court in Philadelphia, a ruling that the U.S. Supreme Court upheld in June 1997.[14] In Europe, where free speech is not given as much constitutional protection as in the United States, attempts to ban certain controversial (especially violent) content on the Internet may be more successful. At a minimum, the European Commission has recommended that member states of the European Union cooperate to ensure that what is illegal off the Internet is also illegal on it, and at the same time to develop minimum European standards as to what types of content should be subject to criminal penalties. Of greater concern to civil liberties advocates, European political leaders also are reportedly looking at tagging or tracer technologies that reveal the identities of

---

14. The Court let stand the Act's prohibition of "obscene" content, a ban that it has upheld in non-Internet contexts.

Internet users as a way of monitoring and then controlling their access to websites government officials believe may be pornographic.[15]

The varied experiences of efforts to control content on the Internet lead to various, even diametrically opposed, pessimistic scenarios. Those in the United States who were disappointed by the Supreme Court's ruling on the CDA fear the societal consequences of the easy access to pornographic and other undesirable content that the Internet makes possible. Indeed, many Americans may be uncomfortable with the fact that the earliest and most profitable commercial use of the Internet has been for X-rated materials (which also was true for the videotape rental business). Now that limits of the CDA on the availability of pornography have been struck down, concern about undesirable content on the Internet has only grown. Civil libertarians fear the very opposite—that any successful attempts to limit content on the Internet, here or in Europe, will lead down a slippery slope toward government censorship. We discuss possible, albeit not perfect, technological solutions to both sets of concerns later.

Digital pessimism about content, however, extends beyond controversial material. China restricts transmission of business information over the Internet. France is considering restrictions that would mandate that all commercial communications on the Internet coming into France be written in French. Some countries sharply regulate the amount and type of television advertising, which acts as an impediment to exports from other countries (including the United States). And a number of countries—such as Australia, Canada, members of the European Union, and Mexico—impose domestic content requirements on their television programs in the interest of protecting their cultures.

For the most part, restrictions of this sort impede commercial activity across national borders and thus represent the kinds of nontariff barriers to trade that are typically subjects of bilateral and multilateral negotiations. In this area, and in others that we discuss in our concluding section, there is an important role for our government to persuade other countries to undo their artificial content restrictions. In addition, government continues to have

---

15. Westin (1996, pp. 19–20).

an important role in prosecuting those who use the Internet to disseminate fraudulent information about products, services, and investments. Because the parties engaged in such activities may be located outside the United States, our government will need to cooperate with other national governments to locate and apprehend them.

Information Overload

Another widely heard complaint about the digital age is that it has spawned the explosion of too much information. Even if they are not "spammed" with junk e-mail, many workers are said to suffer from e-mail overload because they are forced to spend additional hours at work or at home reading and answering their electronic messages. The proliferation of pagers, cell phones, faxes, and other hardware of the digital age has tethered many workers to their offices throughout the day and indeed throughout the year (vacations no longer insulate many from work-related demands). Add the growing number of television channels and magazines to which the average American has access, let alone the likelihood of 500-channel cable television on the horizon, and the result seems to be nothing less than an information avalanche.

One consequence is a rising level of stress reportedly felt by many Americans. There is nothing much government can do about that, but again, technology itself may come to the rescue if there is a sufficient market demand for it to do so. For those plagued with too many e-mails, for example, filtering programs undoubtedly will improve over time to ensure that users get and read only what they truly want to see. For those feeling bombarded by too much information, new services will develop to act as a personal agent to hunt out and bring to an individual each day (or more frequently if so desired) data, news, and commentary on subjects that individual truly cares about. And for those feeling besieged by too many choices offered by television, there is the option of just turning it off or taking an oath of abstinence from the remote control.

Moreover, it is likely that for every American who feels overly stressed by the digital age, there is at least another who welcomes the vast new opportunities for communication and information gathering that the various digital technologies have made possible. Relatives and friends who may have talked infrequently over the telephone are staying in touch on a more regular basis, in much cheaper fashion, through e-mail. Schoolchildren are tapping the seemingly infinite sources of information available on the Internet to do their coursework, to help with homework, and to carry out research assignments. And increasing numbers of Americans and wired individuals around the world—there are actually more Internet users per person in Finland than in the United States, for example—are taking advantage of the Internet to save time and money in searching for products and services.

Nonetheless, it is this richness of choices that has given rise to yet another complaint about the Internet in particular, but also in general about the expansion of television channels: that it is contributing to a fragmentation of society into narrow interest groups, members of which have less interest in the affairs of their community, state, or nation. In the old analog world, Americans got their news primarily from the three major television networks (or local affiliates) and local newspapers, which in combination, it is said, unnoticeably forged a common bond among citizens because they were all exposed to pretty much the same menu of information. In the digital world of seemingly infinite choices, that common bond arguably has been broken, increasingly replaced by a series of other bonds, more narrowly defined along professional, recreational, or family lines.

To the extent that limited choices imposed a common bond on Americans, it certainly was an artificial one, and also not one without criticism. Robert Putnam of Harvard University has pointed to television as the most important culprit in corroding "civil society"—the informal network of civic and religious associations that arguably were once much more important than they are now—and leading Americans to "bowl alone."[16] In any event, the Internet will, over time, continue to erode the influence of television, empowering individuals to become their own broadcasters of information to masses of people. Indeed, by enabling users to forge new bonds with others of like interests, the Internet

---

16. Putnam (1995).

holds the greatest promise for reversing the trends about which Putnam has expressed so much concern. [17]

Decline of the City

Because they substitute for direct personal contact, all improvements in communications and transportation technologies inherently promote decentralization and thus contribute to urban sprawl. The growing use of computers and e-mail, in particular, has enabled increasing numbers of Americans to telecommute, now estimated at 11 million jobs nationwide, up from 4 million in 1990.[18] As computer and communications technologies grow more powerful, some fear that telecommuting will increasingly hollow out the urban core, whose main function now is to host workers during the day but whose only function tomorrow may be to house those with few skills and low incomes (and, in many cases, no jobs).

This concern is hard to dismiss and cannot be disproved or validated without more empirical experience. Nonetheless, there are several factors pointing in other directions.

For one thing, there are likely to be inherent limits to telecommuting. Although some jobs can be performed without regular face-to-face contact and social interaction with co-workers, most jobs will continue to require on-location presence. This is true not only for manufacturing jobs but for most services and retailing. In addition, many firms that have permitted telecommuting have discovered its limits: workers who do not have adequate child care can be less than fully productive, and job performance of telecommuters can be difficult to monitor.

Another consideration is that, in the face of this continuing march of digital technology, a number of major urban centers— New York, Los Angeles, Cleveland, and Chicago, among others—

---

17. A recent *Wall Street Journal* poll highlights the dramatic decline in television viewing by Americans: 65 percent say they are watching less than five years ago; 21 percent pointed to their use of personal computers and on-line services as a reason. Ellen Graham, "Where Have All Viewers Gone," *Wall Street Journal*, June 26, 1997, p. R1.

18. Grimsley (1997).

have displayed a new vitality in recent years. Violent crime rates are down in many urban centers, and some improvements (albeit modest) are being recorded in public education. If further progress can be made on these fronts, then urban America has an increasing chance to attract back some of the middle-class families who have deserted the central core for the suburbs, notwithstanding the forces of telecommuting that may work in the other direction.

Another caveat to the pessimists' concerns is that high technology itself is creating jobs in urban areas. The most prominent example may be the rapid growth of the multimedia industries in Los Angeles, New York, and San Francisco. Another promising sign is the emergence in some cities (Washington, D.C., being a notable example) of innovative efforts to provide computer training (primarily data entry and word processing) to low-income residents of publicly owned or assisted housing developments who can then perform computer-related tasks right where they live. This form of urban core telecommuting could help offset some of the long-distance telecommuting engaged in by individuals now residing in suburban or rural locations.

And finally, the effects of digital revolution are likely to be liberating for many, even if they have some negative impacts on urban areas. In technical terms, the ease and faster pace of communications reduce the economies of agglomeration by increasing the potential for people to interact across space. More people will chose a home based on where they want to live rather than the location of their employer or a satisfactory school. This will spread out the population in the direction of preferred living environments. Many people will choose to interact in specialized communities defined more by common interests and preferences than by geographic proximity. The population, property values, and congestion in many major cities could well decline. The urban population will increasingly consist of those for whom personal interaction is especially important in their work and those who especially value social interaction.

Disappearing Taxes

Yet another claim about the digital age—and the prospect of a growing volume of commerce conducted on-line—is that it will eventually undermine the ability of governments to collect taxes. The problem is neatly spelled out by the following scenario outlined by The *Economist*: "Suppose a customer in California downloads software bought from a firm in Seattle. The company transmits it using the Internet from a computer in Texas. Which state should tax the profit? Or say a German consumer buys a software package from a local subsidiary of an American firm. If he goes into a shop, the profit is taxed in Germany. But if he downloads the software over the Internet, lower, American rates apply."[19]

Courts within the United States and elsewhere will no doubt wrestle with the jurisdictional problems created by the Internet. In the meantime, companies doing business will surely do their best to find ways of arbitraging tax differentials across countries (and to a lesser extent across states within the United States), ideally avoiding taxes altogether. Both income and sales tax revenues could fall victim to this process.

The United States and most other countries tax income on the basis of both where the income is earned (its source) and the residence of the person or entity earning the income. To avoid double taxation of foreign residents in particular, the United States currently has comprehensive tax treaties with forty-eight countries that generally give the residence country an unlimited right to tax income while restricting or even eliminating the source country's right to tax. As Internet commerce grows, it will become increasingly difficult (as the example demonstrates) to determine where income is actually earned. This has led the U.S. Treasury Department to conclude that taxation must increasingly be based on residence (U.S. Dept. of Treasury 1996). Yet this should provide little comfort to governments, because on-line businesses will then have strong incentives to reside in countries with low rates of income taxation.

The message is little different for countries (or states, for that matter) that rely heavily on sales-based taxes, such as those in Europe with its value added tax (VAT). Many states in this country, for example, exempt mail order firms from their sales taxes on sales to nonresidents. As on-line sales grow, more commerce

---

19. The *Economist*, May 31, 1997, p. 22.

could slip through this exemption. In theory, European countries impose VAT taxes even on goods imported from abroad (and so differ from the way most U.S. states treat mail order sales). But for many smaller consumer items ordered over the Internet and shipped into the country in small boxes, let alone software downloaded directly onto the hard disks of Europeans sitting in front of their computers, it is virtually impossible for European authorities to collect the tax.

Other features of digital commerce will compound tax collection difficulties. Various forms of electronic money are now under development or even on the market, including general purpose stored-value or smart cards (for face-to-face transactions and eventually for Internet payment when inserted into a computer port) and different means of transferring money directly over the Internet. For tax purposes, forms of electronic money that have some sort of paper record will allow transactions to be verified and thus pose no new collection problems. But versions of electronic money that allow anonymous transfers (such as the Mondex card, which permits money to be transferred directly between cards without an intervening third party) make it easy to avoid taxes. This problem will grow if consumers use the Internet to open accounts and deposit monies in off-shore accounts where secrecy laws inhibit tax authorities from other countries from auditing transactions.

The coming pressures on tax collections as a result of electronic commerce reinforce trends already under way as a result of globalization, or the increasing integration of trade and investment across borders, facilitated by the enhanced communication capabilities of the digital age. Already, these developments have enabled multinational firms to reduce their tax liabilities by shifting operations and using artful transfer prices on transactions between affiliates in different countries to transfer profits to low-tax jurisdictions. As a recent survey on the subject has pointed out, this process has reduced the relative burden of taxation on highly mobile capital in industrialized countries and increased it on immobile labor.[20]

---

20. The *Economist*, May 31, 1997.

Nonetheless, it is easy to get carried away with predictions about the imminent demise of countries' abilities to tax. There has yet to be a mass exodus, after all, of firms from high-tax countries in Europe to lower tax regions of the world, often where labor is much cheaper as well. This is because firms and individuals take account not only of tax burdens but a whole host of other factors—including the attachment to their friends, their culture, the range of entertainment and business opportunities offered at home, and the amount and quality of public services their jurisdictions provide. Inertia also exerts a powerful bias toward remaining in the same location. If taxes were all that mattered, Bill Gates and Warren Buffett—America's two richest individuals—probably would have moved to the Cayman Islands or some other tax haven long ago.

Still, as costs of communication and transportation continue to fall, an increasing number of firms and highly mobile (and highly skilled) individuals will face rising incentives at the margin to move away from high-tax jurisdictions that are not providing a compensating level of public services. For reasons already outlined, electronic commerce also will pose a growing and eventually significant threat to the tax bases of many countries as well. This will gradually intensify pressure on all governments intent on keeping expenditures in line with revenues to search for the most cost-effective ways of delivering services and to eliminate funding of unnecessary programs and subsidies, developments we both welcome. It is also likely to shift the tax base from income to consumption, as some sources of income become more difficult to monitor.

Governments, however, are not likely to passively accept these developments. Within national governments, there will be increasing demands for centralization of fiscal and regulatory decisions. Across national governments, there will be increasing demands for harmonization of fiscal and regulatory decisions. Governments are increasingly willing to allow private firms to compete across national borders but characteristically oppose competition among governments. For example, this is the underlying rationale for the developing pressure for strong side agreements on labor and environmental standards as part of any new trade agreement.

Digitally Induced Unemployment

Yet another concern about the digital revolution is that it is destroying jobs, replacing people with machines. Some critics, such as Jeremy Rifkin, have forecast the "end of work," with a large fraction of the population having no work to do at all.[21]

This is hardly a new criticism. Seemingly every decade prophets of one sort or another have issued warnings that automation threatens employment. And to a limited extent, of course, they are right. Automation and changes in workplace organization are the keys to rising labor productivity—or steady increases in the amount of goods and services produced per person or hour of labor, which is how nations improve their standards of living. In many cases, firms can maintain or even increase their job rolls even as they become more productive, reducing their prices and thus selling more units. But for firms that are left behind, using older, more expensive technologies or selling older products, sales may decline as their competitors race ahead. In these firms, employment shrinks. This Schumpeterian process of creative destruction is a hallmark of all rapidly advancing and creative economies.

The critical question, of course, is whether the people displaced by this process end up finding new jobs elsewhere in the economy or whether the total number of jobs falls. The U.S. economy at least has consistently rebounded with replacement jobs. Although our economy has experienced its cyclical ups and downs, it continues to demonstrate an impressive ability to generate new jobs despite continued productivity improvements. Indeed, at this writing, the aggregate unemployment rate is hovering at the lowest level in the past quarter century. As a result, even downsized workers have been able to obtain reemployment.

To be sure, many workers who lose their jobs suffer wage losses, as the unique experiences and value they may have built up in the jobs they lose are of little use to their new employers. In addition, firms continue to demand higher skilled employees, especially those equipped with computer skills, which has contributed to widening wage and income inequality. Nonetheless, the ma-

---

21. Rifkin (1995).

jority of jobs generated in the most recent expansion have been "good jobs" paying above-average wages.[22] This result would not be expected if the digital pessimists who have forecast the "end of work" were right.

But what about the future? How can we be certain that as computers continue to replace people, unemployment will not permanently increase? Although nothing is certain in life, there are some things about the economy that have been demonstrated time and time again, and there is no reason to believe that these same truths will not continue to hold in the future. One such proposition is that as some parts of the economy experience productivity improvements, the prices of the products and services produced in those sectors decline. That enables consumers either to increase their purchases, thus maintaining employment in those sectors; or to redirect their purchases elsewhere, where employment must therefore increase. To take a concrete example, if the price of telephone services falls, some consumers will use their telephones more frequently. But others will pocket the savings and spend them on other things, such as on better food, entertainment, recreation, or vacations. As spending shifts among sectors, so does employment. That is why over the past several years, employment increasingly has moved toward health care, travel, and tourism and away from manufacturing industries, whose output continues to increase, but because of healthy productivity improvements, whose need for workers falls.

A different sort of employment pessimism looks beyond our borders and fears that as the workforces of other nations become more educated, especially those in developing countries where wages are low, firms now doing business in the United States will be driven to move their production off-shore. Special attention is given to one of the digital industries—software programming— where in such countries as India software employment has been rapidly expanding.

But worldwide employment in any particular sector is not a zero-sum game. It is not true that an additional job created outside the United States must take away one here, especially in an industry such as software, whose markets continue to expand and

---

22. Council of Economic Advisors (1996).

thus whose employment needs continue to grow. Any quick pe-
rusal of the help wanted advertisements in the newspapers of any
metropolitan area will reveal that workers in the United States
skilled in programming and using computers are very much in
demand, despite the growing numbers of workers with similar
skills outside this country who are gaining high-technology employment.

This is not to deny that some U.S. firms have already moved
some of their computer-related jobs requiring relatively few skills
(such as data entry and elementary software coding) to off-shore
locations, a trend that would seem to confirm the fears of the digital
pessimists. But this pattern is consistent with the way many, if
not most, other industries have developed. They begin in highly
developed countries (such as the United States), and as the firms
in them mature, they routinize their production processes so that
workers with lesser skills can perform them. Because the United
States has long had a comparative advantage in activities requiring highly skilled workers, the jobs with fewer skills gradually
migrate to other low-cost locations, either inside or outside this
country. Nonetheless, the United States manages to keep its labor
force nearly (or actually) fully employed because the cultural, legal, financial, and regulatory climate facilitates the formation of
new businesses and the development of new products and services that continue to require workers to produce and deliver them.
Contrast this experience with Western Europe, where for more
than a decade rigid rules relating to the hiring and firing of workers and a relatively poorly developed capital market (outside the
banking system) have kept employment flat and the unemployment rate roughly twice that of the United States.

The digital pessimists are right in one respect, however. The
steady growth of educated workers abroad reinforces the continuing need for U.S. workers to upgrade their skills if they wish
to improve their incomes. Experience has demonstrated that the
most effective training is provided on the job. Likewise, employers are most likely to provide such training if tight labor markets
effectively force them to do so. Sound macroeconomic policy that
keeps the unemployment rate as low as possible without leading
to the acceleration of inflation is therefore perhaps the best means
of ensuring that those with the least skill and thus the lowest in-

comes in our society are afforded the most expansive opportunities for improving their incomes and standard of living.

Electronic Redlining

To this point, we have considered complaints that in one fashion or another center on the excessively rapid growth and diffusion of digital technologies. The final complaint about the digital revolution we discuss here runs in the opposite direction: that the technologies, especially access to computers and the Internet, are so vitally important, but still expensive, that they are likely to be out of the reach of lower-income Americans. Such electronic redlining, it has been argued, will aggravate any tendencies toward income inequality that we have touched on.

Congress has already responded to these fears by requiring the Federal Communications Commission to create a new universal service mechanism as part of the Telecommunications Reform Act of 1996. Under the previous system, telephone service for low-income Americans and those living in high-cost, rural areas was cross-subsidized through the rate structure in each state. The system was financed by above-cost rates paid by business customers and by long-distance telephone providers, who paid access fees to local telephone companies well above the costs of connection. Similarly, regulators have compelled business customers as a class to subsidize the class of residential telephone users.

The new system, since announced by the FCC, urges the states to move away from cross-subsidies in their rate structures and modestly lowers access charges (although not yet close to cost). In their place, all providers of telecommunications services are required to assess a universal service charge on all customers, with the proceeds (approximately $4 billion per year) used to fund both the long-standing beneficiaries of the universal service system— low-income customers and those in rural areas—and a new program for expanding access to the Internet and other digital technologies to schools that otherwise might not have access to them.[23]

___

23. In particular, the FCC announced in May 1997 a plan to provide $2.25 billion a year in discounts on Internet hookup charges to schools, with the

The new telephone pricing system moves in the direction of supporting a fully competitive telecommunications marketplace. The previous system of cross-subsidies embedded in the telephone rate structure cannot be long sustained now that local telephone markets have been opened to competition, because new entrants will engage in cream-skimming—in other words, serving only the customers (business users) that have been forced to subsidize other users and leaving the incumbent telephone companies that retain universal obligations with a steadily shrinking customer base to support the subsidies.

Nonetheless, the FCC has yet to fully remove the cross-subsidies because of its fear that, in a system in which prices are truly tied to costs, the prices of local residential service in many locations will go up. In fact, as part of its access fee ruling, the commission explicitly capped local residential rates and instead raised business rates (and residences with multiple telephone lines) to help compensate local telephone companies for the loss of some of their access charge revenues. As a result, the pricing of telephone rates is still not consistent with the promotion of a truly competitive market.

More broadly, the euphemism that telecommunications providers are being assessed a universal service charge under the new system of subsidies should be dropped. In substance, the universal service charge is a special tax that is assessed not by Congress but instead by federal regulators. In a similar fashion, the uses to which the telecommunications tax revenues are put are also not being decided by Congress, which ordinarily makes such decisions, but again by regulators. This practice is difficult to defend not only because it skirts the normal budget process, but also because the tax assessed only on telecommunications services distorts the marketplace. It is fair to say that food is even more essential than telephone service, yet we do not impose a special tax on agricultural or other food products to finance the food stamp program, which is meant to ensure universal access

---

largest discounts aimed at schools with high percentages of students from low-income families. During the same month, the Educational Testing Service released a study showing that schools with the highest percentage of minority students had the fewest computers.

to food. Instead, food stamps are funded out of the general revenues the federal government collects (primarily from income taxes). There is little justification for treating telephone, let alone advanced telecommunications services, any differently. If society judges Internet access in schools to be essential, then Congress ought to pay for it directly through the appropriations process, so that the spending on the Internet can be balanced against other priorities (and that the decisions on what is spent are made by elected officials rather than appointed regulators).

Meanwhile, over the longer run, the emergence of competition in local telephone markets coupled with continued advances in wireless technologies should reduce the cost of telecommunications services and thus eliminate any justification for a special universal service program (assuming that such a program could have been justified in the past). In fact, many rural customers already use wireless services in place of land-line telephones. As satellite-based communications systems develop further, many more rural customers will find them affordable, clearly removing any argument for continuing to subsidize much higher cost land-line service. But even if we are too optimistic about the pace of technological advance and the rate at which it will bring down the cost of telecommunications services, if a social judgment is made that telephone services of low-income or rural customers should be subsidized, then the subsidies should be explicit and flow through the budget, rather than continue to distort telephone rates.

Finally, although there is a strong case for ensuring that all schools are wired to the Internet and have adequate computers available for students' use, it is important for policymakers and parents not to lose sight of the more important need for many schools to improve their teaching of the basics—of math, reading, and writing—that are just as essential in the digital age as they always have been. It is not clear that having computers wired to the Internet will contribute to this objective, although once students have mastered basic skills, digital technologies can open new educational horizons. Just as important, in the digital age, knowing how to use a computer and navigate around the Internet will be required skills for operating in adult society. Precisely for this reason, it is only appropriate that the funding and spending for the effort be taken out of the regulatory closet and considered in the open sunlight of the normal budget process. Those who say that a program of wiring the schools and ensuring that they have sufficient computers could not be funded in light of the caps on overall discretionary spending that the Clinton administration and Congress have negotiated are implicitly confessing a lack of confidence in the merits of this type of spending. Moreover, the lack of confidence seems misplaced, especially when as part of the same overall budget agreement, Congress and the administration have agreed both to increases in federal education spending and to new tax breaks for college expenses.

The Productivity Paradox

One important puzzle must be sorted out: the past few decades have been witness to a rapid growth in the development, investment, and use of information technology. Over this same period, however, the rate of growth of (measured) productivity has declined almost continuously, and the variance of earnings by education has increased substantially. If the digital revolution is so important, why is it not reflected in the productivity data? If the digital revolution is so valuable, why have the real earnings of low-skilled workers declined? Or, as MIT economist Robert Solow has quipped, "You can see the computer age everywhere but in the productivity statistics." [24]

There are several answers to this puzzle. One answer is that our economic data are misleading. Our measures of output growth are biased downward for the same reasons that our measures of inflation are biased upward—inadequate measures of the value of new products, quality improvements, and product and outlet substitution. The Boskin Commission, for example, estimated that the Consumer Price Index now overstates the inflation rate by about 1.1 percentage point. Although this estimate has been the subject of some dispute, there is widespread agreement among economists that some amount of inflation overstatement has occurred, which means that real earnings of Americans have been

---

24. Solow (1987).

rising faster than reported by the current statistics and that low-income workers in particular may not be suffering as large an erosion in their real earnings as has been commonly believed.

One issue the Boskin Commission did not address is whether the inflation bias has increased over time, which some observers believe has occurred. If, however, the bias has been roughly constant over the past few decades, actual productivity growth has been higher than the measured rate but has declined by the same amount. In any case, whether or not the real earnings of low-skilled workers have increased or declined, the variance of earnings has increased.

A more comprehensive answer is that slower growth of measured productivity and a higher variance of earnings is characteristic of the early stages of a major technological change. Careful studies of the industrial revolutions in Britain and the United States find that measured productivity growth declined and the variance of earnings increased for the first twenty to forty years after the introduction of the steam engine and, later, electrical power.[25] Only later did major productivity benefits show up as firms and individuals learned how to harness the new technologies to make old things in new ways and to come up with new things entirely. To be sure, the microprocessor has now been with us for a quarter century, but firms and individuals are still learning to adapt to it and, as we suggested earlier, many of its benefits probably lie somewhere in the future.

The two hypotheses that best fit this evidence have interesting implications for understanding recent experience: measured output understates actual output because it does not include the substantial investment in learning how to use the new technology. And educated workers have an advantage in implementing the new technology, but the relative demand for educated workers declines as the stock of learning increases. All of this suggests that measured U.S. productivity growth should increase and the variance of earnings should decline, probably in the next decade. The puzzling combination of current economic conditions, in summary, is not a basis for skepticism about the digital revolution.

---

25. Greenwood (1997).

Solow's skeptical remark also concentrates attention only on computer software and hardware, which still represent a small share of the nation's overall investment, rather than on the broader list of products, services, and technologies—such as telecommunications and broadcasting services and equipment—we believe are associated with the digital revolution and where productivity gains have been more pronounced.

4

Market Solutions

OUR PREFERENCE, as has been stated previously, is to let markets work—that is, to allow firms and individuals to make their own decisions about what to produce and how, and what to buy and at what price—rather than to have a central authority make those decisions. Markets provide incentives for firms not only to produce efficiently but to find cheaper ways of making existing products and to develop new products and services, all because those who do so can get richer in the process. If there was any doubt about the virtue of markets, it was removed when the Berlin Wall came down and exposed the technological backwardness and low productivity of the countries that were once behind it.

Nonetheless, markets are not always perfect. There can be shortcomings even in a market economy that can, at least in theory, justify some sort of government intervention. Perhaps the best known market failures are externalities, both negative and positive. A negative externality occurs when the activity of one firm or individual harms another, pollution being the classic case. In such instances, unconstrained markets will produce too much of the offending good or service. A positive externality, in contrast, occurs when an activity generates benefits that spill over to others, research and development being a prime example. In these cases, markets will produce too little of the activity. Monopoly can be another type of market failure. If not easily contested by other firms, as has long been the case of local telecommunications and electricity transmission and distribution, then unregulated monopolies will charge consumers excessive prices and cause distortions throughout the rest of the economy.

The existence of a market failure, however, does not necessarily justify government intervention. Governments can fail, too— in the way they regulate, subsidize, or otherwise intervene. For example, regulators can set prices too high or too low in their attempts to constrain the exercise of monopoly power or impose social regulations in ways that generate greater costs than benefits. Government interventions can also be ineffective and thus generate unintended distortions. The effort to regulate deposit interest rates of banks and thrifts during the high-inflation environment of the 1970s and early 1980s only encouraged the formation of money market funds that drained depositories of much of their funds (and eventually forced Congress to remove the interest rate ceilings).

The authors of this book represent institutions that often have had different perspectives on the frequency and severity of market failures and on the effectiveness of government in being able to cure them. But we agree on the following principle: government should not intervene in markets without clear evidence that a failure exists; and government action can effectively correct the failure without discouraging market-based solutions to the failure or generating costs that outweigh the benefits of intervention in the process.

Broadly speaking, digital skeptics might argue that the various impediments to the growth of electronic commerce already outlined represent market failures that should cry out for some kind of government intervention. We believe any such rush to judgment, however, would be premature because private actors have strong market incentives to develop technologies to overcome each of the obstacles. Indeed, because so many different digital technologies must work together or not at all (hardware, operating systems software, and applications software), the digital age tends to be characterized by winner-take-all standards, meaning that huge pots of gold await those who can develop satisfactory solutions to concerns about security, privacy, and undesirable content, in particular.

Nonetheless, even if some digital market failures exist, two characteristics of the digital age call into question any assumption that government can or should easily fix the problems, without at the same time causing more harm than good. One obvious feature of the digital revolution is its dynamism, perhaps best illustrated by the operation of Moore's Law. Named after Intel co-founder Gordon Moore, this law posits that the computing power of microprocessors doubles every eighteen months, making possible new applications every time it does. Anyone who has tried to buy a computer, for example, notices that the industry is constantly coming out with more powerful models, followed closely by new and more powerful software.[1]

The constantly changing nature of digital technologies means that government intervention in the digital field runs high risks not only of being premature—a technological solution may be quickly developed that is far more effective than any regulatory edict—but also of frustrating further innovation if the intervention is misplaced or falls victim to the law of unintended consequences (as so many government interventions do). Indeed, government intervention is often sought or supported by particular interests who may not be representative of the broader social interest. The decades of government regulation of prices and entry in all facets of the transportation industry—few of which are characterized by natural monopoly—are but one illustration of this tendency. Moreover, government action is typically responsive to particular crises and thus inherently backward looking. This mindset is not well suited to the digital marketplace, in which constant change is perhaps the only constant.

A related feature of the digital age is that the same technologies that are shrinking time and space—the microprocessor, the Internet, satellites, and fiber-optic cables, to name a few—also are rendering it increasingly difficult for government to enforce its regulations or other interventions because the subjects can increasingly flee to other jurisdictions. As we have already observed, attempts to regulate content on the Internet by controlling what servers may transmit, for example, can be easily circumvented by

---

1. In September 1997 Intel unveiled a new type of computer chip—code-named "StrataFlash"—that the company claims will cut the eighteen months of Moore's Law in half.

moving the servers elsewhere. Similar problems confront any government intent on placing special taxes on transactions completed on the Internet.

In short, the presumption favoring market—rather than governmental—solutions to digital problems arguably is stronger than in other spheres of economic or social activity. Indeed, the march of digital technology poses a strong challenge to many existing rules written for the older analog age, when regulations were more easily enforced—a subject we turn to in the next section.

But we first consider how markets already are addressing or may be likely to tackle three of the impediments or drawbacks to growing Internet-based activity—security, privacy, and pornography—and why market-based solutions may be superior to government mandates or regulation. A word of caution is in order, however, before we begin: none of the solutions—whether driven by the market or by the government—eliminate all of the risks associated with each problem. This should not be surprising or unduly troubling. Virtually everything we do carries with it some risk. Among the critical questions that we address are whether markets can be counted on to reduce certain digital risks to an acceptable level; and if not, whether government action is capable of achieving that objective without creating other significant harms in the process.

Security

Because hackers who steal vital secrets and even money can cause significant damage to any enterprise, all organizations that store or transmit data have a strong interest in having the best data security systems available. The market, therefore, provides strong incentives for firms to meet this demand. In fact, firms now routinely employ encryption techniques that are available on the market or that have been developed in-house to safeguard their own data. Government agencies use similar techniques to protect sensitive information.

Yet as we outlined in the previous section, digital skeptics question whether the market will be able to provide sufficient security—or perhaps more important, the appearance of sufficient security—to attract substantially more users to make their purchases over the Internet. For reasons we will spell out shortly, the skeptics' case has been bolstered by the various attempts by the Clinton administration to restrict the exportation of encryption technology. But first, some quick history is in order.

In 1993 the Clinton administration announced that it would allow the exportation of "strong" encryption technologies (at the time anything with keys longer than forty bits) only if they incorporated the Clipper chip, or computer circuitry that gave "spare" keys to two government agencies acting as "escrow agents" (the Treasury and Commerce departments). The administration's policy had a laudable objective: to ensure that either law enforcement or national security authorities could, under appropriate circumstances, gain access to the keys to decode encrypted messages in order to prevent criminals and terrorists from using advanced cryptography to escape detection.

The Clipper chip proposal generated a firestorm of protest from industry and foreign governments who did not trust any governmental organization (even two separate agencies) to hold the keys. In addition, critics claimed that the policy only benefited foreign producers that were beginning to develop and market to customers in this country and elsewhere around the world encryption technologies that were stronger than what U.S. companies could export. And Clipper "key escrow" offered no business value to users—spare keys were available only to the government and not to users who had lost their own keys through mishap or accident.

The administration responded by changing its policy. In late 1996, it announced that it would unconditionally allow exportation of 56-bit encryption; however, stronger encryption could be exported only if U.S. exporters demonstrate that they have viable means to permit the recovery of encryption keys or plain text for use by law enforcement officials, under legal authorization. The new policy permitted individuals within user organizations to serve as internal "key recovery agents" provided they stood ready to hand over the keys to the authorities (with a court order) and would not notify surveilled parties within the same company that they were being watched or investigated. Great Britain and France have endorsed the key recovery concept, but Germany and Israel (among other countries) remain opposed.[2]

Most of the criticism of key recovery so far has focused on its threat to individuals' privacy and the potential harm to the U.S. software industry. A collection of leading computer scientists signed a statement in May 1997 suggesting that the approach was technologically flawed and could not work at reasonable cost.[3] A broader concern is that, as with the Clipper chip, foreign producers not subject to any key recovery system will continue to have an advantage in marketing their systems relative to U.S. companies. This places continued pressure on both the private sector and the government to find ways of circumventing any mandatory, uniform key recovery requirement, which means that it either will not be as effective as advertised or will be undermined by the very authorities who are determined to administer it.

For example, one private way to circumvent an escrow requirement is for U.S. firms to sell from foreign locations encryption technology developed abroad. In fact, Sun Microsystems announced such an arrangement in May 1997 with its plan to sell technology developed by a foreign supplier (in this case from Russia) in which Sun had a minority interest—a plan the Commerce Department is contesting but may not be able to stop. As for the government, even the Clinton administration has excepted encryption technology used by financial institutions from the key recovery requirement in order to facilitate electronic commerce (and in June 1997 approved the exportation by Microsoft and Netscape of electronic banking software that contains powerful 128-bit keys).[4] And then there is the easiest way of all to evade

---

2. The FBI has taken a harder stance on the encryption issue. Director Louis Freeh has urged that all encryption software, whether sold here or exported abroad, should contain a feature allowing law enforcement agencies to decode encrypted messages. Although sophisticated computer users could disable this feature, the presumption would be that relatively few would do so.

3. Abelson and others (1997).

4. The "financial institutions exception" may be of less value than it first appeared because all parties to financial transaction—not just the financial institution—must be allowed to use the encrypted software. But the department's exception is not this broad.

any government restrictions—to carry software out of the country on a diskette or by supplying it from servers overseas directly on the Internet.

Given all of the objections and concerns raised about a mandatory key escrow system, we would suggest that, at most, the government require exported encryption technology to contain features enabling (but not forcing) users to use key recovery (which, in turn, would allow authorities to obtain the keys under appropriate legal circumstances).[5] Many, if not most users would choose such a feature in their own interest to ensure that they could recover their own keys should they lose or forget them. If the government required that the option be made available for export purposes then software firms would have incentives to produce software containing the feature, but domestic users would not be mandated to use it.

This is probably the best that can be done to meet the legitimate needs of the law enforcement and national security authorities. Yet even this compromise policy may not be tenable in the long run because of the incentives for programmers outside the United States to develop and market strong cryptography that may not contain key recovery (although foreign programmers may be influenced by or connected with their home country intelligence agencies, which could limit the attractiveness of their cryptography).

Meanwhile, more stringent export restrictions than those just outlined run a major risk of being counterproductive—that is, discouraging the use and development of strong encryption technologies that would help allay skeptics' concerns about security in digital communications.[6] In a global economy, multinational

---

5. Technology has been developed to limit access of law enforcement to particular documents or sessions rather than to all the files that may be on a user's disks (hard or floppy).

6. At this writing, Congress is considering two alternative approaches for weakening the export restrictions. In the House, Representative Bob Goodlatte has proposed a bill that would prohibit mandatory key escrow and abolish export controls on commercial encryption. In the Senate, a bill cosponsored by Senators Robert Kerrey and John McCain would relax some existing restrictions but require people sending encrypted information to the federal government to use key recovery. In addition, the bill would make

firms must be able to communicate in a secure fashion with their foreign suppliers, customers, and business partners. As Kenneth Dam has pointed out, it generally is not cost-effective for companies to maintain two different encryption systems for their communications—a stronger one for their internal domestic communications and a weaker one they are allowed to export for messages sent to entities abroad. As a result, export restrictions tend to drive companies to use weaker cryptography across the board.[7]

One final point is worth noting. We suspect that given the strong market incentives for developing more secure means of electronic payment, especially for consumer transactions over the Internet, the security problems will be resolved to the sufficient satisfaction of most potential users. If we are right, then electronic commerce is likely to grow very rapidly in the years ahead. But even if it does not grow rapidly, the Internet will continue to provide the important and increasing benefit of enabling consumers to locate suitable products and services at the lowest possible cost. As more useful and powerful search engines are developed, the price differentials for the same or similar products and services should considerably narrow, if not vanish, bringing about the frictionless capitalism forecast by Bill Gates in his *The Road Ahead*. Consumers thus stand to benefit in substantial fashion from the search capabilities offered by the Internet whether or not they feel sufficiently secure to use it to pay for their purchases.

Nonetheless, as to payments, one relatively recent set of developments illustrates in striking fashion how the very real prob-

---

using encryption to commit a crime a separately punishable offense (in order to meet concerns in the law enforcement community about encryption without key recovery).

7. Dam (1996). Critics of key recovery also argue that export-same restrictions discourage innovation by suppliers of encryption technology as well, who find it expensive to develop two versions of their software. In order to avoid sacrificing their export sales, many software firms may therefore develop and market the weaker, exportable encryption systems. Although many popular software programs exist in different versions and in different languages, this is not the case for programs produced by smaller firms for whom the cost of developing different versions for domestic and export markets would be significant.

lems that consumers may identify can be resolved by markets rather than government mandates. Although government intervention originally helped jump start the general-purpose credit card industry by limiting cardholders' liability to $50 for lost cards and fraudulent charges, it has proved unnecessary for the newer debit cards, which many consumers complained carried unlimited liability. During the summer of 1997, MasterCard was the first to respond to consumers' concerns by imposing a fifty-dollar liability on its own cards, which Visa then one-upped several weeks later, not only by matching the MasterCard policy but by announcing that it would eliminate all consumer liability on unauthorized credit and debit card purchases if consumers reported the problems within two business days. Cynics may charge that both organizations moved because of the impending threat of legislation mandating similar liability limits on both debit and credit cards. That may or may not be true. But the cynical view cannot explain the competition between the two organizations over the liability policy that ensued, illustrating the virtues of market-driven solutions.

Certification and Authentication

Related to the issue of security is the question of reputation and trust—factors that are as important for electronic commerce as any market. Customers will want assurance that a website is, in fact, sponsored by a real business and not an electronic imposter. Electronic commerce also will require authentication services to ensure that electronic documents are not forged or altered. Contrary to the belief in some quarters, government is not required to provide certification and authentication functions: both can be supplied by the private sector.

Public-key encryption will play a major role in achieving this objective. Public-key cryptography enables the recipient of a message to identify its sender using a digital signature. The sender encrypts part of the message—the signature—with his or her private key, which only each user knows, and at the time enciphers the message with the recipient's public key. The recipient decrypts the message with his or her private key and confirms the sender's identity by verifying the signature with the sender's public key.

When both the recipient and the sender are using public-key technology, encryption can provide privacy (a subject we discuss later), security, and authentication. Certification requires a trusted third party to attest that a certain public key really belongs to a certain individual or business. Companies are already providing such services.

New legislation is not necessary to protect certification authorities from liability in the event that someone fraudulently obtains or uses a certificate of identity because certification authorities can insulate themselves from liability by contract. We would abandon our sanguine view in this regard, however, if courts did not respect parties' freedom to contract in this manner.

It is important to recognize that many electronic transactions will not require certification. Most businesses do not need to know the identity of their customers, only that they will be paid. Anonymous digital cash, for example, provides business with this assurance without certification of personal identities (although it can complicate the lives of those charged with law enforcement).

There may be a role for the federal government, however, to ensure that the laws governing what constitutes an acceptable digital signature do not impede the growth of electronic commerce. At this writing, many states already have enacted laws in this area, but they are not consistent with one another, nor are they all technologically neutral—that is, some of these laws single out only specific technologies as acceptable. This emerging Tower of Babble appears to be in need of a federal guiding hand, one that would allow parties engaging in electronic commerce to adhere to either a federal standard for digital signature (one that is technologically neutral) or the relevant state standard.

Privacy

Can unfettered markets be counted on to meet the apparently growing concerns about privacy in an electronic world, or is some kind of government regulation needed? The case for markets rests primarily on technology's being able to give users the ability to choose how much information to provide to vendors or other potential recipients of data over the Internet. In fact, a number of labeling technologies are already in the process of development or on the market. For example, the Open Profiling Standard (OPS), developed by a group of software companies and proposed to the World Wide Web Consortium, which develops standards for the Web, would allow users to specify what information they want to reveal to any particular website and have stored on their hard drives (so, for example, a user could ask that his or her name and e-mail address be provided but no other personal information). OPS stands in stark contrast to the current "cookies" that many websites now routinely plant in users' computers, generally without users' knowledge, to greet them the next time they visit and to enable the originating websites to track other sites users may be visiting (information that can be sold to advertisers and other interested parties). OPS will enable users effectively to disable their cookies, a task that can be done already by using software that can be downloaded for no cost from the Internet.

Consumers also may want to know what merchants do with the data they collect. To satisfy this concern, an industry consortium named TRUSTe has been formed to provide the electronic equivalent of the Good Housekeeping Seal of Approval to websites that maintain confidentiality of their data. TRUSTe will audit the licensees of its logo to ensure that they are adhering to their announced privacy policies.[8] The American Institute of Certified Public Accountants, working in conjunction with its Canadian counterpart at Verisign (a digital certification company), has announced a similar service that will rate a website's data security, integrity, and privacy.

These are only the first examples of what is likely to be a continuing stream of technologies that will be developed to empower users to choose how much privacy protection they actually want. Privacy itself will then become a commodity, with each user deciding how much privacy he or she may be willing to pay for: Those who value it highly will refuse to do business with mer-

---

8. TRUSTe will provide three "trustmarks" to help inform consumers about the privacy practices of websites: "No exchange" (indicating that no personal information is used); "One-to-One Exchange" (indicating that the data collected is only for the use of the site owner); and "Third-Party Exchange" (indicating that the data are provided to third parties but only with the consumer's knowledge).

chants who do not follow a scrupulous policy of protecting confidential information, whereas those who have lesser concern for privacy will do business with other merchants, at perhaps some cost saving. Moreover, systems such as OPS also will help shield those computer users who provide only limited information about themselves to other sites from junk e-mail (inhibiting direct marketers and other database companies from targeting them as potential consumers of particular products and services).

And then there is the weight of public opinion, which in an electronic age can easily and promptly manifest itself on literally any subject, including privacy. Thus, when America Online (AOL) announced in July 1997 that it was planning to sell its members' telephone numbers to third parties, it was so deluged with objections by e-mail that it withdrew the policy the next day. The Experian episode discussed earlier produced a similar outcome. These events demonstrate how in yet another fashion technology makes it possible for markets to meet objections of digital skeptics or pessimists.

Nonetheless, critics have attacked the private sector approach on a number of grounds, alleging that:

—Many consumers may have only a few merchants from which to choose and thus cannot effectively bargain to preserve their privacy if the merchants insist on asking for the information.

—Providing choices to children is meaningless for children, who are easy prey for websites asking for information in seemingly innocuous ways.

—Even if users can effectively limit the information they provide, they now have no way of correcting wrong information that may be compiled about them in computer files, as they do with their credit histories because of legislation (the Fair Credit Reporting Act).

It is likely that these reasons contribute to the finding of at least one recent opinion poll that a majority of Internet users favor the enactment of some kind of Internet privacy law.[9] Some privacy advocates would model such a law on the comprehensive approach followed by the European Union and thus require collec-

---

9. *Wall Street Journal*, June 19, 1997, p. B6.

tors of data to notify users about what data is being collected about them and for what purpose; to obtain consent from users (preferably affirmatively, or an "opt in" basis, rather than negatively, or "opt out"); and to afford users meaningful opportunities to examine and correct their files ("access"). If the EU model were followed, then the United States would create a federal agency to enforce these requirements.[10]

Given the dynamic nature of digital technologies, especially the techniques that recently have been developed for empowering potential data subjects to choose their own level of privacy, we believe comprehensive federal privacy legislation fashioned on the EU model is highly premature. At least one national survey also shows the concept to be unpopular among members of the public, although it is not clear how well many of the respondents actually understood the issues involved.[11]

To those who claim that Internet merchants have unequal bargaining power and thus may not give consumers a choice about providing personal information, we note simply that consumers can still purchase products through more conventional means. To those who would urge comprehensive privacy legislation in the name of "protecting children," we would point to the difficulty, if not impossibility, of crafting any law that is not overly broad. How, after all, is it possible to ascertain whether a particular site is aimed especially at children? Moreover, how can regulations of those sites be enforced in a world in which servers can simply move off shore and continue business as usual? At some point, users—and, if not them, then their parents—must take responsibility for their

---

10. The United States currently has a patchwork of state and federal statutes that protect privacy. At the federal level, the most prominent examples include the Federal Privacy Act of 1974 (which limits the collection of data by the government on citizens while giving the right to discover, correct, and control dissemination of information about them in the government's possession, but has no parallel provisions for business); the protection of videotape rental records; and the Right to Privacy Act of 1978, which protects against disclosure to the government of personal financial records held at banks, except with a search warrant. At this writing, Congress is also considering legislation to protect the confidentiality of medical records.

11. Westin (1996).

actions. In any event, technologies such as OPS that allow users to specify on their hard drives exactly what information about them can be revealed to other websites should solve the problem by allowing parents automatically to restrict the ability of their children to provide information to third parties.

There are two possible areas in which government intervention may be appropriate, although not without problems. One is narrowly crafted legislation that would require entities offering personal information on the Internet to afford the subjects opportunities to examine and correct their entries. We admit, however, that it is difficult to see how such a requirement can be enforced (the off-shore problem again) or how consumers can be made aware of the myriad databases that may be storing information about them. Another possibility is for the government to require or encourage Internet merchants to allow consumers to make purchases anonymously, as Germany has done. Here, too, enforcement would be difficult. Moreover, goods still must be delivered to an address, so that their receipt cannot be anonymous.

Any legislation aimed at bolstering privacy runs other dangers. Databases and those who "mine" them provide useful services: enabling law enforcement authorities to locate criminal suspects and witnesses, locate abducted children, and find parents delinquent in paying child support; permitting financial institutions and merchants to reduce fraud losses, which are passed on to consumers in higher prices of products and services; facilitating the verification of information on mortgage and other credit applications; and enabling companies to market products and services selectively to the most likely prospects, thereby reducing the volume of junk mail. Overly intrusive requirements meant to stop privacy abuses can unwittingly deprive many others of the benefits a now-open Internet provides.

Finally, privacy policy in the highly dynamic Internet age must take account of the comparative virtues and disadvantages of governments and markets. Legislation is prone to force on actors one-size-fits-all requirements that can stifle change. Regulation, meanwhile, takes time to develop and can always be subject to lengthy judicial challenge. Markets can and do move much more swiftly and deftly, once practices are brought to light. And while government intervention may appear to be more effective than markets in ridding cyberspace of undesirable privacy practices, the supposed enforcement advantage of the government looks much less substantial when servers can be moved easily to other jurisdictions with less stringent laws. In the digital age, consumers ultimately will get the privacy they want if they demand it.

Controversial Content

Just as technology can and will empower individuals to choose how much information about themselves to reveal to others on the Internet, technology already has made it possible for individuals—especially pertinent to parents—to choose what kinds of content, if any, they would not like to see transmitted to them or their children. In particular, various commercially available software programs and some Internet service providers (such as America Online) either filter out objectionable material or allow users to choose a level of material to which they prefer to be exposed. In addition, the World Wide Web Consortium is developing a labeling standard called Platform for Internet Content Selection (PICS), to be completed some time in 1998, that creates a way to both rate and block on-line content. PICS will allow different private rating organizations to devise their own rating schemes from which users can choose.

To be sure, there are limits to all filtering technologies; none is perfect, as demonstrated in a test administered by *Consumer Reports* and reported in its May 1997 issue. Some perfectly legitimate material may be unintentionally blocked, whereas some cleverly disguised material may nevertheless slip through. Still, the various filtering technologies remove virtually all of the risk of unwelcome material's reaching children. Any remaining portion could be eliminated if parents closely monitored their children while on the Internet (by doing something as simple as locating the family computer in common areas of the house). And if all else fails, parents can (and should in any event) warn and educate their children ahead of time that there are things on the Internet that are inappropriate for them, just as there are in the nonvirtual world.

Filtering software and labeling standards—PICS in particular—nonetheless have attracted criticism in some libertarian circles.[12] The main charge: that the technology provides powerful new tools to countries, such as Singapore and China, to censor information rather than having simply to block access to a broad class of data. In effect, the complaint is that filtering provides the censorship equivalent of a rifle rather than a shotgun and thus tempts governments to engage in even more censorship than they otherwise might.

We are more sanguine, for two reasons. For one thing, there is little danger of censorship in the United States, as the Supreme Court's clear striking down of the Communications Decency Act attests. To be sure, PICs and other filtering devices make it easier for schools to screen information made available to children on the Internet, but this is no different in substance than the current policy schools have toward printed materials. As for governments that are likely to take advantage of PICS, these are governments that would attempt to block information in any event, only much more of it without PICS than with it. In short, filtering software is broadly empowering and enables individuals to make more informed choices than they otherwise could. That is an outcome that free market advocates should applaud rather than denigrate.

---

12. Lessig (1997).

5

The Proper Role
for Government

WE HAVE JUST REVIEWED how the market is likely to resolve most of the perceived problems of the digital revolution. We now suggest what the government should do. At the risk of some repetition, it is useful to summarize the principles that should guide this division of roles.

To begin with, we should take our ignorance more seriously— a point made most forcefully by Nobel laureate F. A. Hayek and emphasized by Peter Pitsch at the Brookings-Cato conference. The information marketplace will develop in ways that cannot be anticipated and should not be forced. The market is a superior discovery process—especially under conditions of great uncertainty—less likely to make big mistakes, and quicker to correct small mistakes.

This perspective is not uniformly shared. Robert Kuttner, a leading economic journalist, wrote, "American industry is in one of its great periods of innovation and recombination. At such moments, elite opinion wants to sweep aside all concerns for a broader public interest that industry, in its myopia, cannot pursue. But it is precisely at times like these that government is needed to steer and stabilize."[1]

---

1. Robert Kuttner, "Clinton's Talented and Tenacious Regulators," *Washington Post*, June 2, 1997, p. A19.

We respectfully disagree. It is precisely at times like these that government does not have sufficient information to steer intelligently and would make a major mistake to stabilize any specific set of conditions. So far, fortunately, the Clinton administration seems to share our position on this issue.

Second, government, with the best possible information and motives, will make some mistakes, but it should not repeat the major mistakes of the past. Most major prior mistakes in government-industry relations have been a coincidence of bureaucratic imperialism and a competitive threat to large private firms.

About a century ago, for example, the inventions of Alexander Bell, Thomas Edison, Guglielmo Marconi, and others created the products and industries that we now call the second industrial revolution. By the late 1920s, after the original patents had expired, the dominant firms in the telephone, electricity, and broadcasting industries faced the prospect, and in some cases the reality, of considerable competition. Yet the dominant firms in these industries sought regulation, primarily to thwart that new competition. As Peter Huber has noted in the case of telephone regulation,

> By the late 1920s, support had begun to emerge for sweeping federal control of the telephone industry. Monopoly telephone service had become so familiar it seemed inevitable. Several ponderous studies officially confirmed that it was a conclusion perfectly consonant with the New Deal political winds then blowing. Bell itself wanted to consolidate its dominant position and legitimize its monopoly. Theodore Vail spoke publicly in favor of regulation, sounding a theme ("cream skimming") that would become a Bell rallying cry for the next half-century. "If there is to be state control and regulation," Vail argued, "there should also be state protection to a corporation striving to serve the whole community ... from aggressive competition which covers only that part which is profitable."[2]

Samuel Insull, head of the National Electric Lighting Cartel, made the same argument—supporting regulation of electricity in exchange for exclusive regional monopolies. The broadcasters, who had earlier developed a de facto system of property rights in

---

2. Huber (1994, p. 267).

the frequency spectrum, bowed to strong pressure to cede these rights to the federal government in exchange for restrictions on new entrants. Again, in Peter Huber's acid commentary, "Private stations would broadcast at the commission's pleasure, like peasants tending their cattle on the pastures of the crown."[3]

One older industry faced a similar problem in the 1920s. Railroads had been loosely regulated beginning with the Interstate Commerce Act of 1887, an act that was primarily designed to prevent the type of price discrimination that is necessary to finance any good or service with declining average costs. In the 1920s, however, railroads began to face substantial competition from unregulated trucks. And the railroads, alas, also became strong advocates of rate and entry regulation in exchange for broadening this regulation to their new competition.

In each case, the dominant firms in these industries made a Faustian bargain with the state, promoting or accepting a massive increase in regulation in exchange for the easier life of a regulated monopolist. Seventy years later, we have yet to unwind many of these unnecessary and inefficient restrictions on our economy.

We provide this history as a warning of a similar scenario that is very important to avoid. Any new technology, any new good or service, any aggressive new competition threatens the net worth and easy life of some incumbent firm. And someone in the government is usually looking for some new reason to regulate. The new technologies of the digital age promise a third industrial revolution but present the same type of threat to some interests and the potential for a whole new wave of regulation.

Third, realizing the full potential of the information marketplace will present special challenges for government policy. Once a reasonable degree of actual and effective potential competition takes hold throughout the telecommunications industry, there will no longer be a viable basis for regulating the transmission of voice, data, and video by separate standards, or indeed for separately regulating the industry at all. The most effective role for government in the meantime will be to help ensure that this competition comes about.

With these principles in mind, we now address the major types

---

3. Huber (1994, p. 267).

of policy issues that will affect the development of the information marketplace. How will digital technology challenge existing policy? And how are these issues likely to be best resolved?

Antitrust

The information marketplace presents a number of special challenges to conventional antitrust policy. In particular:

—The proliferation of somewhat different goods and services makes it difficult to define the relevant market. A company may have a temporary monopoly in some niche market without having any significant enduring monopoly power.

—Some important digital goods and services are even more valuable to each user the more other people use them, a characteristic economists describe as a network externality. This has long been true, of course, for telephone service, which has grown more useful, if not essential, as more people are hooked up. Some of the technologies that define the digital age also have this characteristic: operating systems for personal computers, formats for audio- and videotape recording, and formats for digital broadcasting of television signals. In these cases, the technology becomes a standard on which other technologies depend. And standards, in turn, almost inherently tend to be monopolies.

—Some cooperative action, especially joint ventures among otherwise competitive firms, is often the most efficient way to select standards for interoperability or to provide network services. Such joint ventures, however, risk antitrust attention and the arbitrary application of anticartel rules.

—The specialized resource in the digital age is human creativity, not physical capital. There is no way to prevent an effective merger that is achieved by hiring the key personnel of a major competitor. On the other hand, there is no way to maintain an effective monopoly of the key personnel against the opportunities to work in another firm.

—Finally, many developments in the information marketplace happen very quickly compared with the glacial speed of the characteristic antitrust case.

How should the antitrust authorities respond to these challenges? As a general rule, be cautious. There is no reason to develop an antitrust policy specific to the information marketplace, but there is reason to question how traditional antitrust rules developed in a nondigital world should apply in the digital age.

First, and most important, a firm's success should not be a sufficient basis for antitrust attention. A producer that attains a monopoly status by legal means that benefit consumers and do not foreclose the opportunities of other entrants should not be subject to an antitrust violation, in the language of Judge Learned Hand in the 1945 Alcoa case, "merely by virtue of his superior skill, foresight, and industry."

Second, the objective of antitrust enforcement should be to ensure, to the extent possible, that all markets, including those that may be currently characterized by monopoly, are contestable and that existing firms can be challenged by firms offering new or lower priced products or services. A related implication is that the antitrust authorities should focus on barriers to entry, not on market shares. Thus antitrust authorities must accept the fact that the specialized digital markets may be characterized by monopoly for some period of time, as long as there are no unlawful barriers to entry by another product or service. There is a valuable continuing role for antitrust to ensure that these markets remain contestable, such as in the 1994 consent decree in which Microsoft agreed with the Department of Justice to abandon a number of licensing practices that protected the company's dominance of the market for operating systems.[4] In addition, the authorities will need to remain vigilant to ensure that no dominant player in any digital market discriminates against competitors in much the same way that, almost two decades earlier, AT&T was found to have done with respect to long-distance telephone providers (who needed access to AT&T's local loop monopolies).

But the antitrust authorities also should recognize that their

---

4. In October 1997, the Justice Department charged Microsoft with violating the terms of the decree by allegedly requiring computer manufacturers to license Microsoft's new Internet browser, Internet Explorer 4.0, as an integral part of the company's operating system software. In December the District Court for the District of Colubia issued a preliminary injunction against this practice, which Microsoft appealed.

power to influence fast-moving technology markets is inherently limited. Indeed, Netscape's browser and the growing importance of Java software for the Internet almost surely have done more to limit Microsoft's dominance of operating systems than the 1994 consent decree. Moreover, even Microsoft has not been able to leverage its dominance in the operating systems market to control the on-line service market—where AOL instead has remained far in the lead—as many feared it would when it introduced Microsoft Network.

Third, modern antitrust should eschew its traditional suspicion of joint ventures, at least in the digital realm. A special problem exists in the information marketplace when no one firm has sufficient incentive to establish a network standard or service. In this case, the alternative means to establish a network standard or service are a joint venture by private primary market firms or for the government to select the standard or provide the service. The antitrust authorities, however, have often subjected joint ventures to a compulsory access rule, permitting a primary market competitor to force its way into a successful network.[5] Meanwhile, governments do not have a good track record in selecting network standards—witness the abortive attempt by the European Union and Japan to set an analog standard for high-definition television—and they generally impose some regulatory conditions as the price of access to such government network services as the Federal Reserve check-clearing system.

Again, we urge the government to let the market sort out these issues. In some cases, such as the competition between the Beta and VHS formats for videocassettes, consumers without any collective action by a joint venture or governments selected the preferred standard. More recently, the FCC wisely decided not to mandate a standard for HDTV, allowing manufactures to offer competing technologies.

In some cases, however, a joint venture may be the most efficient type of institution to select a network standard or provide a network service. The government would be wise to immunize such joint ventures from antitrust attention, as long as the network standards and services themselves are open to competition

---

5. Baker (1994).

from other existing or potential networks. Starting in the 1980s, Congress approved a similar policy affecting joint ventures for export promotion, for research and development, and for production. And the European Union has adopted a similar policy affecting joint ventures of several types. As summarized by Donald Baker,

> Antitrust policy should always favor competition based on efficiency (as the Supreme Court explained in 1975); and, where efficient network alternatives are possible, it should favor competitive networks. Therefore antitrust law should require a rigorous initial analysis of the consequences or monopoly power of the Network facility; based on its cost characteristics and the lack of competitive alternatives to it, before application of any compulsory access rule. No time is perfect to rethink awkward legal legacies from the distant past—but this is a good time to try, because network competition can be a vital part of our post-industrial future.[6]

Domain Names

The developing controversy about the type and number of top-level domain names (TLDs)—the suffixes that are part of every e-mail or website address on the Internet—is best resolved by clarifying the relevant legal rights and full privatization of these rights.[7]

The central problem of this system is that no organization has a clear legal authority to determine the type and number of TLDs or to establish rules and procedures for the root-name servers (the computers that contain the directories used by the domain name system). Since 1989 TLDs have been managed by IANA, a government contractor operating out of the University of Southern California's Information Sciences Institute. IANA, however, does not have the authority to assign the relevant legal rights to competing private firms or to manage the root-name servers owned by other organizations. Such a legal vacuum does not encourage self-government; it perpetuates confusion and invites the Internet

---

6. Baker (1994, p. 60).
7. We draw in this section on the arguments set forth in Mueller (1997).

equivalent of land grabs and squatting.

The U.S. government must take the lead to establish the legal and institutional basis for private sector competition and ongoing self-governance. The best model for a private domain name system is probably the system of company symbols adopted by the stock exchanges, where each exchange actively manages the symbols used in that exchange and there is substantial competition among the exchanges. The World Trade Organization, in turn, may be the best forum to establish the rules for privatization of the global name space.

A competitive market in domain name registration would probably add 250 to 350 more TLDs. This is the best way to permit a broader selection of preferred individual addresses and to avoid the misleading and unnecessary controversies about trademark rights. In twenty years something quite different from domain names may be used to locate information on the Internet. In the meantime, domain names should be allowed to evolve as much like natural language as possible.

Content Control

Content control—censorship—is perhaps the most contentious early policy issue affecting the information marketplace. We have already addressed how we believe the market will allow individuals to exercise choice—with regard to what they want to view and what they want their children to view—without government intervention. Nonetheless, because issues relating to content control in the digital age are likely to continue to be of some interest, we reemphasize here what we believe governments in particular should and should not do.

The principles suggested by Ithiel de Sola Pool in his classic 1983 book *Technologies of Freedom* are as relevant today as they were when he wrote them more than a decade ago:

—Governments should impose no prior restraint on Internet content. This may not be possible in any case if the origin of the communication cannot be identified or the type of communication is legal in that jurisdiction.

—The smallest relevant group of adults should have full au- ·

thority to determine what content comes off the Internet, probably assisted by available software filters. This principle would permit content control at the level of each household, firm, school board and library board but not by either the individuals within these organizations or by a larger general government. The available software filters are not perfect but are likely to improve in response to the market demands for content control.

—All legal rules affecting communication should be uniform across the means of communication. Any communication that is legal on other media, thus, would be legal on the Internet. And those who initiate communications on the Internet would be subject to the same penalties for libel, slander, extortion, fraud, copyright infringement, breech of proprietary or government secrets, conspiracy to commit a crime, and so forth as on other media. For too long, the FCC has set content standards on radio and television that are different from the standards that apply to speech and the print media. The rapidly increasing use of the Internet should provide the opportunity to eliminate all the media-specific rules on the content of communications.[8]

Admittedly, these principles are not sufficient to satisfy those who, for whatever purpose, would restrict the types of communications available to other adults. They should be sufficient to protect individuals from unwanted content—whether in the form of pornography, violence, advertising, or unsolicited e-mail.

Licensing

Individual states have the authority to license occupations, and hundreds of occupations—from practicing medicine to braiding hair—are licensed somewhere in the United States. These licenses have long restricted the supply of labor in these occupations and have limited the innovation and division of labor in the relevant markets. Until recently, there was not much of an issue about the potential for the division of labor across state borders. Services were provided by the local licensed practitioner, and there was

---

8. Pool (1983).

little opportunity to make use of specialists in other states.

The Internet, however, creates opportunities for a wide range of teleservices, such as telelegal services and telemedicine, but these opportunities are seriously restricted by state licensing laws. For example, these laws allow the sale of legal programs prepared in another state but ban legal services on the Internet. Activities now legal, such as the use of mail or the telephone, to consult with a specialist in another state, are illegal on the Internet. Information from every medical test can be provided on the Internet, but this opportunity is caught in a legal web whether the patient or the doctor has moved to make these tests possible. Some states allow free telemedicine services but ban such services for a fee. And so on.

The potential for mutual recognition of those licensed in other states to practice law is restricted by the significant differences among some types of state law. In this case, there may still be an opportunity for reciprocal licensing among a group of states or for the separation of a general license to practice law from a certificate identifying a specialty in the laws of a specific state or body of law. It is also important to avoid similar state restrictions on classroom instruction over the Internet in the name of protecting the dubious value of state teaching certificates.

Most state laws that license physicians date from the 1870s, predating both the automobile and the telephone. It is clearly the time to bring these laws up to date. All new physicians now take the same national exam of more than 2,000 questions. That should be a sufficient basis for a mutual recognition of the medical licenses granted in another state or the replacement of state licenses with a national license. The federal government has the clear authority to regulate interstate commerce, and this may be a case of whether a federal rule is superior to the combination of state rules.

Beyond our borders, the Internet is allowing providers of goods and services to market their wares not just in this country but throughout the world. What happens, for example, when America's Internet-based discount brokers offer to trade stock to citizens of other countries? Securities regulators in Great Britain have been grappling with that question and at least so far have insisted that advertisements of banking and brokerage services on the Internet must comply with British advertising laws (which require advance approval of print-based advertising copy of such services). It is not difficult to imagine how this kind of thinking can get out of hand, and not just for these particular financial services. The logic that says that simply because something is available on the Internet must also mean that it must comply with the national laws of each country where citizens may have access to the Internet is an Orwellian logic that could quickly cripple electronic commerce. A far better method is for countries to adopt an approach of mutual recognition whereby regulators from countries in which services may originate certify compliance and allow (if not mandate) that fact to be advertised on the website where the service is offered.[9]

Product Standards

For the most part, the government should stay out of the business of setting product standards. Setting a standard often requires a dominant firm to accept substantial risk or a group decision by otherwise competing firms. The market process for setting standards is often messy, sometimes leads to false starts, may require a joint venture, is especially difficult when the agreement of two industries is required—and usually works. Good examples of standards set by the market include those for videocassettes, cellular telephones, direct broadcast satellites, fax machines, and e-mail. The often-repeated stories about the market being locked into early inferior standards have not proven to be correct, but the potential for "path dependence" still exists.

The government, however, is a blind giant with neither the information nor the incentive to choose a superior standard. As we suggested in our antitrust discussion, one valuable action would be to immunize joint ventures organized to set a standard from antitrust attention as long as the initial membership in this joint venture is open to any firm. The May 1997 Mutual Recognition Agreement by the U.S. government and the European Union to accept each other's testing standards for a wide range of products (including appliances, pharmaceuticals, and telecommuni-

9. "Beware the Cyber-Regulator," The *Economist*, August 23, 1997, pp. 56–57.

cations equipment) is also valuable and should be broadened to include other products and other governments. Again, the appropriate principle is to facilitate the market rather than to preempt it.

Remaining Regulatory Tasks

The Telecommunications Act of 1996 set the stage for continued deregulation but was also sufficiently open-ended to permit new forms of regulations and cross-subsidies. The potential for continued deregulation will depend on future decisions by both Congress and the FCC. The stakes are enormous. Telecommunications companies, which are busy spending an estimated $40 billion a year in this country alone to upgrade their networks, are attempting to wrestle with the uncertainties wrought by the rapid changes in the legal and technological environments—and specifically the rapid movement from carrying voice to data. Indeed, in the years ahead virtually all of the growth in telecommunications revenues here will come from data rather than voice traffic.[10]

Especially critical will be the resolution of clear and fair rules for access to the local telephone exchanges. Current charges imposed on long distance carriers to connect to local exchanges contain huge cross-subsidies, as we have already discussed in connection with electronic redlining and universal service. The long distance carriers are correct in highlighting this issue. At this writing, the FCC continues to struggle with this problem, while trying to find a way to finance new cross-subsidies to provide Internet access for schools and libraries—all without increasing local service rates.

Meanwhile, a contentious debate rages over the terms under which companies seeking to enter the local telecommunications business (including, but not limited to, the long-distance carriers) can connect to the regional local exchange carriers that now essentially have monopolies in this business. This issue is now before the Supreme Court. As a matter of economics, however, the principles for resolving this dispute should be clear: Those who want access to elements of the local telephone network should

---

10. Kupfer (1997).

be charged their long-run incremental cost, which should include a fair (risk-adjusted) return on future capital projects in order to encourage new investment.

Regardless of the outcome in the courts and the agencies, digital technology may force a resolution of these issues. The Internet will reduce the potential for cross-subsidy, depending in part on the rapidly improving quality of digital telephony, because over time more users will make their calls over the Internet rather than through conventional means. To be sure, serving the Internet will require substantial additional investment by the local exchanges, but the potential access fees are limited by the cost of several competing technologies to provide Internet access.

It is not just the United States that will be wrestling with these issues. As we noted at the outset, in February 1997, the U.S. government and the governments of sixty-eight other members of the World Trade Organization (WTO) agreed to open their national markets for basic telecommunications services to both foreign competitors and foreign investors. Each member government is required to guarantee nondiscriminatory access to the local exchanges and transparent rules for the use of the frequency spectrum. Anticompetitive practices such as cross-subsidies are banned. As a trade agreement, these commitments are subject to WTO dispute-settlement procedures backed up by the potential for rule-based retaliation.

We offer the following suggestions for resolving the access charge issue:

—Substitute direct tax-financed subsidies for all cross-subsidies of politically preferred customers such as rural residents, schools, and libraries.

—Allow the local telephone exchanges to charge congestion fees on all peak-hour users (a practice that long-distance companies have engaged in for years).

—Reduce access fees to long-run incremental cost (including an allowance for return on capital).

These proposals, however, will not avoid controversy. Subsidized customers probably prefer a hidden cross-subsidy rather than one dependent on an annual appropriation. Many customers are quick to complain about both congestion and congestion fees. There will always be disputes about the estimates of long-run marginal cost and what is a fair return on capital. And until local telephone markets are reasonably competitive, the FCC (and state regulators) will have at least one raison d'etre: to arbitrate complex disputes about the extent to which the Bell Operating Companies are complying with the requirements of the 1996 act to facilitate interconnection with competing local telecommunications carriers. But no one promised our regulators and politicians a rose garden.

The other major task now remaining within the FCC's jurisdiction is to create full property rights in the frequency spectrum. This requires that these rights be independent of the status of the owner (broadcaster, small business, minority firm, and so forth), fully transferable to other buyers, well defined (including provisions for what constitutes an acceptable minimum degree of signal interference), and unrestricted by type of use. There is not now (or maybe never was) a need for a regulatory commission to resolve potential interference problems in the use of the spectrum: the courts and the common law of trespass should be sufficient for this purpose if the relevant property rights are adequately defined.

Once these issues are resolved, what is left for the FCC? Shortly after announcing his intention to leave the FCC, Chairman Reed Hundt suggested two new missions:

—solve the campaign finance problem by giving political candidates free television time;
—"export" the FCC to provide advice on the telecommunication policies of other governments.

These suggestions sound like an agency searching for a mission. We do not need an administrative agency, run by nonelected officials, to solve the campaign finance problem, which is inherently a political issue that should be addressed by political leaders. And the only international role we see for the FCC's telecommunications expertise is to assist our trade negotiating office to ensure that other countries live up to their recent agreement to open their telecommunications markets.

One of the more interesting controversies that developed at the Brookings-Cato conference concerned the necessary remaining life of the FCC. Lawrence Gasman of Cato argued that the two major remaining issues—access charges and spectrum—could be resolved in eighteen months. Robert Crandall of Brookings cautioned that it might take five years, especially to resolve the thorny issues involving when and under what conditions the regional Bell Operating Companies can be permitted into long distance.[11] We will not make a specific terminating proposal because the time required for all telephone markets to become reasonably competitive is difficult to predict. Yet now that policymakers seem to have embraced competition as the guiding principle to govern the telecommunications industry, they have also, at least implicitly, laid the intellectual groundwork for the eventual termination of the FCC. The agency's demise, therefore, should not be prolonged beyond the time necessary to give competition a reasonable chance to thrive in all telecommunications markets.

Long live the digital age.

---

11. Crandall (1997).

References

Abelson, Hal, and others. 1997. "The Risks of Key Recovery." *Key Escrow, and Trusted Third-Party Encryption* (May 27).

Baker, Donald I. 1994. "Compelling Access to Network Joint Ventures." *Regulation* 2: 53–60.

Clinton, William J., and Albert Gore Jr. 1997. *A Framework for Global Electronic Commerce.* July.

Council of Economic Advisers. 1996. "Job Creation and Employment Opportunities: The United States Labor Market, 1993–1996." April 23.

Crandall, Robert W. 1997. "Are We Deregulating Telephone Services? Think Again." Brookings Policy Brief (March) no. 13.

Dam, Kenneth W. 1996. "The Role of Private Groups in Public Policy: Cryptography and the National Research Council." Occasional Paper 38. University of Chicago, Law School.

Dam, Kenneth W., and Herbert S. Lin. 1996. *Cryptography's Role in Securing the Information Society.* National Academy Press.

Deutch, John. 1996. Testimony before the Permanent Investigations Subcommittee of the Senate Governmental Affairs Committee (June 25) (responding to questions from Senator Sam Nunn).

Dyson, Esther. Release 1.0 (Newsletter available at http://www.edventure.com/release1/release1.html)

Flamm, Kenneth. 1997. "Deciphering the Cryptography Debate." Brookings Policy Brief 21 (July).

Gates, Bill. 1995. *The Road Ahead.* Viking.

Greenwood, Jeremy. 1997. *The Third Industrial Revolution: Technology, Productivity, and Income Inequality.* AEI.

Grimsley, Kirstin Downey. 1997. "Telecommuting's Growth Marked by Glaring Glitches." *Washington Post* (July 5): 1, 10, 11.

Huber, Peter. 1994. *Orwell's Revenge: The 1984 Palimpsest.* Free Press.

83

Katz, Jon. 1997. "Birth of a Digital Nation." *Wired* (April) (http://www.wired.com/wired/5.04/netizen.html).

Krueger, Alan B. 1993. "How Computers Have Changed the Wage Structure: Evidence from Microdata." *Quarterly Journal of Economics* 108 (February): 33–60.

Kupfer, Andrew. 1997. "Transforming Telecom: The Big Switch." *Fortune* (October 13): 105–18.

Lessig, Lawrence. 1997. "Tyranny in the Infrastructure." *Wired* (July): 96.

Mueller, Milton. 1997. "Internet Domain Names: Privatization, Competition and Freedom of Expression." Cato Briefing Paper 33 (October 16).

Negroponte, Nicholas. 1995. *Being Digital.* Knopf.

Pool, Ithiel de Sola. 1983. *Technologies of Freedom.* Harvard University Press.

Putnam, Robert D. 1995. "Bowling Alone: America's Declining Social Capital." *Journal of Democracy* 6 (January): 65–78.

Quittner, Joshua. 1997. "Invasion of Privacy." *Time* (August 25): 28–35.

Rifkin, Jeremy. 1995. *The End of Work: The Decline of the Global Labor Force and the Dawn of the Post-Market Era.* G. P. Putnam's Sons.

Sichel, Daniel E. 1997. *The Computer Revolution: An Economic Perspective.* Brookings.

Solow, Robert M. 1987. "We'd Better Watch Out." *New York Times Book Review.* July 12, p. 36.

Swire, Peter, and Robert E. Litan. 1998. *None of Your Business: World Data Flows and the European Privacy Directive.* Brookings (forthcoming).

U.S. Department of the Treasury. 1996. "Selected Tax Policy Implications of Global Electronic Commerce." Office of Tax Policy (November).

Westin, Alan. 1996. "Data Protection in the Global Society." Conference Report, American Institute for Contemporary German Studies, Berlin, November 15.

Conference Participants

Does the Digital Age Require a New Approach to Regulation (If So, What Is It)?

Michael L. Katz, University of California, Berkeley
Peter K. Pitsch, Pitsch Communications
Lawrence J. White, New York University
Richard E. Wiley, Wiley, Rein & Fielding

Does the Digital Age Require a New Approach to Antitrust?

Donald I. Baker, Baker & Miller
Timothy F. Bresnahan, Stanford University
Robert E. Hall, Stanford University
Thomas W. Hazlett, University of California, Davis

Making Global Commerce Happen (including, Intellectual Property, Encryption, Privacy, UCC Contracts, Anti-fraud, and Tax Issues)

Anne Branscomb, Harvard University (since deceased)
Kenneth W. Dam, University of Chicago
Dorothy E. Denning, Georgetown University
Whitfield Diffie, Sun Microsystems, Inc.
David Post, Temple University Law School

Professional Licensing

Joel Hyatt, Hyatt Legal Services
Jay H. Sanders, American Telemedicine Association
Robert J. Waters, Arent Fox

Regulating Content (Culture, Pornography, and Libel)

Walter Berns, American Enterprise Institute
Richard D. Klingler, Sidley & Austin
Eli Noam, Columbia University
Patrick Vittet-Phillipe, European Union

Standards and Interoperability

Cynthia Belz, American Enterprise Institute for Public Policy
    Research
Daniel L. Brenner, National Cable Television Association
Jeff Rohlfs, Strategic Policy Research

Ensuring Access

Robert Crandall, the Brookings Institution
Lawrence Gasman, Cato Institute
Jonathan Sallet, MCI
Lawrence Strickling, Ameritech (now at the FCC)

Index

Workplace: benefits of digitization, 3; computer literacy as income factor, 3, 15; decentralization of power in digital age, 2–3; employment patterns, 42–45; global competition for labor, 43–44; information technology investments, 13; management structure, 3–4, 15; productivity paradox, 48–50; professional licensing, 75–77; skills training,

44–45; technological future, 14, 15; technology jobs in urban areas, 38; wage trends, 42–43. *See also* Telecommuting
World Intellectual Property Organization, 28, 29, 30
World Trade Organization, 79; recommendations for policy, 8; regulation of domain names, 74
Wyden, Ron, 26